Dynamics of Robots with Contact Tasks

International Series on

MICROPROCESSOR-BASED AND INTELLIGENT SYSTEMS ENGINEERING

VOLUME 26

Dynamics of Robots with Contact Tasks

by

MIOMIR VUKOBRATOVIC
Robotics Laboratory,
Mihailo Pupin Institute,
Belgrade,
Serbia & Montenegro

VELJKO POTKONJAK
Faculty of Electrical Engineering,
University of Belgrade,
Belgrade,
Serbia & Montenegro

and

VLADIMIR MATIJEVIC
Robotics Laboratory,
Mihailo Pupin Institute,
Belgrade,
Serbia & Montenegro

KLUWER ACADEMIC PUBLISHERS
DORDRECHT / BOSTON / LONDON

A C.I.P. Catalogue record for this book is available from the Library of Congress.

ISBN 978-90-481-6515-5

Published by Kluwer Academic Publishers,
P.O. Box 17, 3300 AA Dordrecht, The Netherlands.

Sold and distributed in North, Central and South America
by Kluwer Academic Publishers,
101 Philip Drive, Norwell, MA 02061, U.S.A.

In all other countries, sold and distributed
by Kluwer Academic Publishers,
P.O. Box 322, 3300 AH Dordrecht, The Netherlands.

Printed on acid-free paper

Printed in the Netherlands.

Contents

Preface

Regarding the nature of interaction between a robot and its environment, robotic applications can be categorized into two classes. The first one covers the non-contact (unconstrained) motion in a free space, without any relevant environmental influence exerted on the robot. In these tasks, the robot's own dynamics has a crucial influence upon its performance. A limited number of most frequent simple robotic tasks, such as pick-and-place, spray painting, and arc welding, belong to this group.

In contrast to these tasks, many complex advanced robotic applications, such as assembly and machining require the manipulator to be mechanically coupled to other objects. In principle, two subclasses of basic contact tasks can be distinguished. The first one covers essential force tasks whose very nature requires an end effector to establish the physical contact with the environment and exert a process-specific force. In general, these tasks require both the position of the end effector and the interaction force to be controlled simultaneously. Typical examples of such tasks are machining processes such as grinding, deburring, polishing, etc. In these tasks the force is an inherent part of the process and plays a decisive role in the task execution (e.g., metal cutting or plastic deformation). In order to prevent overloading or damage of the tool during the operation, this force must be controlled in accordance with some definite task requirements.

The prime emphasis with the tasks of the second subclass lies on the end-effector motion, which has to be realized close to the constrained surfaces (compliant motion). A typical representative of such tasks is the part mating process. The problem of controlling the robot during these tasks is, in principle, the problem of precise positioning. However, due to imperfections inherent to the process, sensing and control system, these tasks are inevitably accompanied by the occurrence of contact with the constrained surfaces, which results in the appearance of reaction forces. The measurement of interaction force provides useful information for error detection and appropriate modification of the prescribed robot motion.

Looking at these problems from a broader point of view, one can conclude that they actually belong to the large family of contact problems involving interactive action of at least two dynamics, one originating from an object, structure, or any other active system, and the second, originating from the constraints – the environment of a dynamic nature. Examples in contemporary technical practice are numerous and diverse, and particularly ubiquitous in robotics, ranging from the interaction between a moving load and the ground, and the problem of underground excavation, meat deboning problems in the food industry, to recent medical robot applications in surgery, such as spine surgery, neurosurgical, and microsurgical operations.

The importance of dynamic interaction between the robot base and the ground has also to be emphasized. In order to evaluate the dynamic interaction, local ground condition around the base should be precisely taken into account. The problem of dynamic contact can also be recognized within the hemo-dynamic system of living organisms. Thus, the fluid flow through collapsible tubes represents a problem of contact of the dynamic environment (arteries and veins) and the moving fluid (blood), possessing its own dynamics.

Contact tasks have promising prospects in the field of technical sciences and various engineering disciplines. They can be clearly discerned, and we believe that the time of their concrete solutions is coming. They are considered in brief in the concluding remarks entitled "New Trends in Contact Tasks". As examples, we have mentioned issues such as the road vehicle and its dynamics in contact with the dynamic environment; the railway vehicle and its dynamics from the pantograph via the vehicle structure to the contact with the rail of dynamic character; humanoid robot in contact with the ground during the walking cycle, from an open to a closed kinematic chain; interaction of the mechanical model of the operator with the platform, subjected to programmed disturbances of different character, and other tasks in which, starting from sufficiently exact simulation of the interaction of the object with its environment, the common goal is to attain dynamic control with both position and force of the interactive systems thus coupled.

The future will certainly hold more tasks for which the interaction of a dynamic system with its environment is fundamental.

A common feature of all the above contact tasks is the presence of the constraints upon the robot's motion due to the environmental objects. Supposing all parameters of the environment and robot be known and robot positioning be ideally precise, it might be possible to accomplish the majority of these tasks using the same control strategies and techniques developed for the control of robot motion in free space. However, in reality, none of these conditions can be fulfilled. Therefore, the contact tasks are characterized by the dynamic interaction between the robot and environment, which often cannot be predicted accurately. The amount of mechanical work exchanged between the robot and environment during contact may in many cases vary drastically, and this causes significant alteration of performance of the robotic control system. Hence, for the successful completion of a contact task either the interaction forces have to be monitored and controlled, or a control concept ensuring the robot's compliant interaction with the environment should be applied.

The book is concerned with the problem of contact dynamics and points out the following: *how to derive dynamic models, how to use them*, and, *how to explain the influence of "new" effects*. The book consists of seven chapters and two appendices.

Chapter 1 titled **"Robot Dynamics – Problems, Research, and Results"**, reviews briefly the following ten key stages in robot dynamics.

Problem 1: *open kinematic chain* (with all robot elements considered non-deformable).

Problem 2: *rigid-body contact* – robot environment is considered in the form of a geometric constraint.

Problem 3: *friction between the robot and the environment.*

Problem 4: *introducing flexible links.*

Problem 5: *elastic deformations of torque transmission.*

Problem 6: elastic effects of the robot support.

Problem 7: elastic effects on the environment side of the robot.

Problem 8: elastodynamics in the contact zone on the environment side of the robot.

Problem 9: analysis of deformation on both sides of contact, environment, and robot.

Problem 10: collision modeling through elastodynamics.

Chapter 2, "Free Motion of a Rigid-Body Robot Arm", considers the problems of dynamics assuming that the robot links are infinitely rigid. The discussion starts with the relevant notions from theory of mechanisms (kinematic pairs and chains, degrees of freedom, etc.) and then defines the parameters that describe the robot link geometry. Kinematics and dynamics concern free motion of the robot arm. Introducing the actuator dynamics and elastodynamics of the transmission system expands dynamic model of the robot chain. Finally, some issues of control are elaborated.

Chapter 3, "Rigid-Body Contact of a Robot with Its Environment", elaborates mathematical formulation of contact tasks (a general approach), and particularly the problem of robot arm subject to surface-type constraint (writing task).

Chapter 4, is titled **"Soft and Elastodynamic Contacts"**. Section 1 considers the problem of soft contact – a simplified approach to modeling contact tasks. The more complex approach considers deformation as a dynamic process and takes into account the local dynamics of particles. This approach is called the elastodynamic contact and is discussed in Section 2, where the following issues are elaborated: mathematical formulation of the general problem, surface-type environment, and the problems of flexible-joint robot and flexible-link robot in contact tasks.

In **Chapter 5, "Robot Interacting with Complex Dynamic Environment"**, we consider two mechanical systems being in contact, paying equal attention to both. This generalization is followed by a more complex treatment of contact deformation. Namely, we cover deformation on both sides of contact, on the robot side and on the environment side. In this chapter, the following issues are elaborated: mathematical formulation of the rigid-body contact, modeling friction effects, introducing contact elastodynamics, and the case study (two examples are presented: one dealing with the contact of a robot and a transport cart, and the other discussing more complex cooperation, the problem of assembly).

In **Chapter 6**, titled **"Dynamics of Multi-Arm Cooperative Robots with Elastic Contacts"**, the procedure of modeling and the complete general form mathematical model of manipulators with six degrees of freedom (DOF) in cooperative work are presented, together with the solution of the indefiniteness problem with respect to force distribution. The obtained model is presented in several convenient forms. For the first time, a system of active spatial six-DOF mechanisms elastically interconnected with the object (dynamic environment) is modeled. The reason for the emergence of the indefiniteness problem with respect to force is explained and the procedure for solving this problem is given. Unlike the approaches given in the available literature, the indefiniteness problem with respect to force is solved in accordance with physical phenomena.

In the concluding discussion, **Chapter 7: "New Trends in Contact Tasks"**, the authors confirm the need for research in contact dynamics and highlight a broader class of technical problems where contact and contact dynamics are the key issues. Some of these problems, which may be unexpected in a robotic context, have already become the topic of applied research and implementation.

Regarding the organization of the presentation, it is necessary to mention an important dynamic effect – **impact** – and explain how it is treated in this book. Its elaboration is not given in one particular section but rather spread throughout the text. Each chapter and each example includes discussion on impact effects.

The fact that dynamics and control of contact tasks are strongly related problems induced the authors to provide a brief description of relevant control issues. Another reason for the discussion on control was to help the reader to better understand the simulation experiments conducted in the book. This is in the two appendices, written in provided condensed forms.

Appendix 1 discusses the concept of ***Hybrid position/force control***.

The ***synthesis of control laws for the robot interacting with dynamic environment*** is presented in **Appendix 2**.

In this way this research monograph obtained its necessary degree of autonomy from the control point of view.

This book is intended for all researchers in applied robotics, robot designers, control engineers, as well as postgraduate students of robotics and active systems in general. The background required includes knowledge of the basics of linear and nonlinear systems theory, as well as a fundamental knowledge of active mechanisms dynamics.

June 2003 The Authors
Belgrade, Serbia and Montenegro

Acknowledgments

The authors are grateful to Professor Goran Djordjevic, from the Faculty of Electrical Engineering, University of Nis, for his help in solving specific robotic task, based on redundant active mechanisms.

We also extend our thanks to Dr. M. Zivanovic. The paper: "General Mathematical Model of Multi-Arm Cooperating Robots with Elastic Interconnection at the Contacts" published in the Trans. of the ASME, *Journal of Dynamic Systems, Measurement and Control*, Vol. 119, pp. 707, 1997, by M. Zivanovic and M. Vukobratovic, served as the exclusive basis for preparing Chapter 6 in this book.

It is our pleasure to thank also Dr. D. Surdilovic for his cooperation with the first author of this book resulting particularly in review papers on contact tasks.

The authors indebted to Professor Luka Bjelica for translating part of the manuscript and editing and proofreading the whole text.

Finally, we would also like to thank Svemir Popic, B. Sc. (Mech. Eng.) for his invaluable technical assistance in the final stage of preparing this monograph.

Chapter 1

ROBOT DYNAMICS – PROBLEMS, RESEARCH, AND RESULTS

We start our discussion on robot dynamics from the standpoint that successful design and control of any system requires qualified knowledge of its behavior. This is certain, but we should explain what is meant by "qualified knowledge". Let us consider a robot as an example of a technical system. Qualified knowledge of its behavior may, but need not, include the mathematical model of its dynamics. In the earlier phases of robotics development, design was not based on exact calculations of robot's dynamics but followed the experience gained in machine design. The control did not take account of many dynamic effects. Large approximations were made to reduce the problem to the well-known theory of automatic control. The undeveloped robot theory could not support a more exact approach. For a long time, the practice of robotics (design, manufacturing, and implementation) was growing independently of the theory which was too academic. However, this did not prevent the manufacturers from constructing many efficient robots.

At the present time, the need for complex, precise, and fast robots requires a close connection between theory and practice. Computer-aided methods for dynamic modeling allow fast and user-friendly calculation of all relevant dynamic effects. In this way dynamic modeling and simulation have become essential tools in robot design and control.

In this chapter we present in a very concise way an overview of problems and results of research in robot dynamics.

First studies in robot dynamics dealt with the robot representing an anthropomorphic mechanism [1]. In the beginning, robotic (locomotion or manipulation) mechanisms, robot links, as well as all transmission elements (shafts, gears, etc.), were considered nondeformable (Problem 1 in Fig. 1.1).

Many authors have worked in this field and we mention only a few of the early works [1–10]. In their approach the researchers covered many important dynamic effects (inertia of robot links, dynamics of actuators, energy loss in the transmission system, etc.) and for a long time discussions on robot dynamics were restricted to such problems. This is still the case with most introductory books and textbooks in robotics. The efforts in this direction have resulted in some practical tools for robot design and robot control. Regarding the application of robot dynamics, the main breakthrough was made when the computer-aided methods for dynamic modeling were developed. Such methods were presented in [2, 4, 5] and further efforts resulted in the commercially available software systems [11, 12]. The methods allowed a fast and user-friendly calculation of all relevant dynamic effects. In this way dynamic modeling and simulation have become essential tools in robot design. It was the first step towards CAD systems for robots [13, 14]. The other possibility for the application of robot dynamics has been the synthesis of the so-called dynamic control [15]. It introduced the feedforward term based on the calculation of robot dynamics.

The fact that many robot tasks involved the contact of the end-effector and the robot environment, led to the first research on contact dynamics [16, 17]. Robot environment was considered in the form of a geometric constraint (Problem 2 in Fig. 1.1). Stationary and nonstationary cases were discussed. The constraints imposed on the end-effector resulted in reaction forces. Dynamic models were derived for arbitrary constraints (restricting one to six degrees of freedom). All robot elements were considered rigid. The derived models enabled solution of motion along with computation of contact forces. The friction between the robot and the environment was included (Problem 3). Collision problems were discussed and the solution for nonelastic impact was found.

Some practical examples considered were the writing and assembly task.

After solving the dynamics of rigid robots, the attention of researchers turned to elastic effects. The problem of flexible links was considered first (Problem 4). Some of the initial research looked for a simplified but fast solution of a general flexible chain [14, 16], while others tried to find a more exact model for some practical examples, i.e. single-link or two-link flexible arms [18, 19]. Almost all existing dynamic models for flexible multi-link arms are formed by some specific discretization methods, such as lumped mass [14, 16], assumed modes [20], or finite elements [21]. A few researchers considered the computational efficiency of proposed procedures [22, 23]. Real time calculation has recently become significant in the advanced control algorithms for flexible robots [24]. Many researchers considered the linear deformation and neglected the effects of coupling between the components of deformation. Novel research results [25, 26] take account of these problems, thus producing more general models.

Calculation of the link deformation allowed the development of an applicative CAD system for robots [16, 27].

The next source of elastic effects is the transmission of torque. It is known that with electrical drives, complex transmission between the motor and joint shaft is usually needed. It is necessary to reduce speed and multiply torque, thus, some kind of gearbox is present. Depending on the design, the gearbox introduces smaller or larger elastic deformations. The deformation is especially expressed with Harmonic-Drive reducers since elasticity is the essential property of their operation. If the robot construction is such that gravitational load is reduced by placing all motors close to the robot base, then a system is needed to transmit torque and motion from the motor to the corresponding joint. This may be a chain, belt, shaft, etc. Any of these systems introduces its elastic deformation (Problem 5 in Fig. 1.1). If the transmission is considered deformable, then the motion of the joint becomes independent of motor motion and only the relatively high stiffness makes these motions close to each other. The number of degrees of freedom (DOFs) is at least doubled. The initial results in this field were presented in [28] and [29] and they provided the foundations for further investigations. Mathematical model was derived to describe the dynamics of robot with elastic transmissions. The torque transmission included Harmonic-Drive reducer, gears, and chains. Work [29] followed from practical work in robot design. Special attention was paid to some practical problems in forming the control loop: should one measure the joint position or the motor angle? Generally speaking, the presence of nonactuated DOFs represented the main problem with control of such robots. Further works have elaborated this subject in more detail. One way of solving the tracking problem was presented in [30] and included the measurement of torque for the formation of the feedback. The next generalization in this research was to introduce the constraints upon the motion of the end effector, and thus, consider elastic joint robots in contact tasks. Several approaches to simultaneous force and position control in constrained robot systems with joint flexibility have been proposed in the literature [31–36].

Elastic effects related to the robot base may be involved (Problem 6 in Fig. 1.1). If connections of the robot arm and its base were considered deformable or if the robot was placed on a platform with pneumatic wheels, then oscillations would appear. However, these effects can be included in the existing models of rigid system, by adding passive DOFs with stiffness and damping. Similar problems may appear on the environment side. If the object that should be grasped, or processed in some other way, is connected to its base by means of deformable connections, then oscillations will arise (Problem 7). The dynamic model of the environment is then needed.

With contact tasks, the most interesting deformation effects are expressed in the vicinity of contact points (Problems 8 and 9). Two bodies in contact produce a force upon each other and the force depends strongly on their elastic properties. For exact modeling of contact, the elastodynamics in the contact zone has to be taken into account. In most investigations the contact deformation was considered on the environment side only (Problem 8). The robot's terminal link was assumed nondeformable. Such an approach can be justified in many industrial applications: it is based on the fact that the tools are generally harder then the objects they are acting upon. A general approach, however, would require the analysis of deformation on both sides of contact, environment, and robot (Problem 9) and such a discussion is needed not only for purely theoretical reasons. With some applications, such as peg-in-hole assembly, it is a real situation. The same material is used for the peg and for the object with the hole. Hence, it is likely that both bodies in contact will undergo deformation. Most research efforts in contact dynamics were made with the aim of creating adequate control strategies for contact task control. Different approaches to the problem solution of contact task control can be distinguished; among these, two approaches should be recognized: hybrid position/force control [37–40] and impedance control [41, 42]. Hybrid control consists of two parallel feedback loops, for the position and for the force. Each of these loops uses separate sensor systems and separate control laws are adopted for each loop. Impedance control represents a strategy for constrained motion rather than a concrete control scheme. The objective of this control concept is to achieve a specific mechanical behavior at the manipulator end effector, referred to as target impedance. A certain drawback of conventional impedance control is its applicability to a limited class of contact tasks. Environment was modeled by appropriate impedance and thus, dynamics was included, but with the restriction to linear model. The complete dynamics of the environment, including nonlinear effects and non-holonomic constraints, is the topic of dynamic position-force control of robots interacting with dynamic environment [43–45].

Special attention should be paid to the problem of collision (Problem 10). It is an omnipresent effect since no contact can be made so precisely to avoid impact. The first study of impact with robotic systems was given in [46]. The nonelastic impact between the robot and object in grasping was solved. In [16], the collision of the robot end-effector and a geometric constraint was elaborated. The impact was still considered plastic. The effect of friction was included. Both of these studies followed the classical approach based on the law of momentum. Another early result is [47]. The influence of friction on the body collision was discussed in detail in [48, 49]. Rigid-body collision of planar kinematic chains with multiple contact points was considered in [50]. A successful algorithm for the numerical integration of the system subject to impact was presented in [51]. In [52], the mathematical model of impulsive collision dynamics was formed through the use of Schwartz's distributions, then, the relationships between impulsive and continuous dynamic models were examined along with the analyis of the difficulties associated with the transition phase control. In [53], the impact was modeled as a "black box", without requiring to explicitly observe the compression and restitution phase. In order to achieve better insight, the collision could be modeled through elastodynamics. One way of doing this was by means of the lumped mass approach [54].

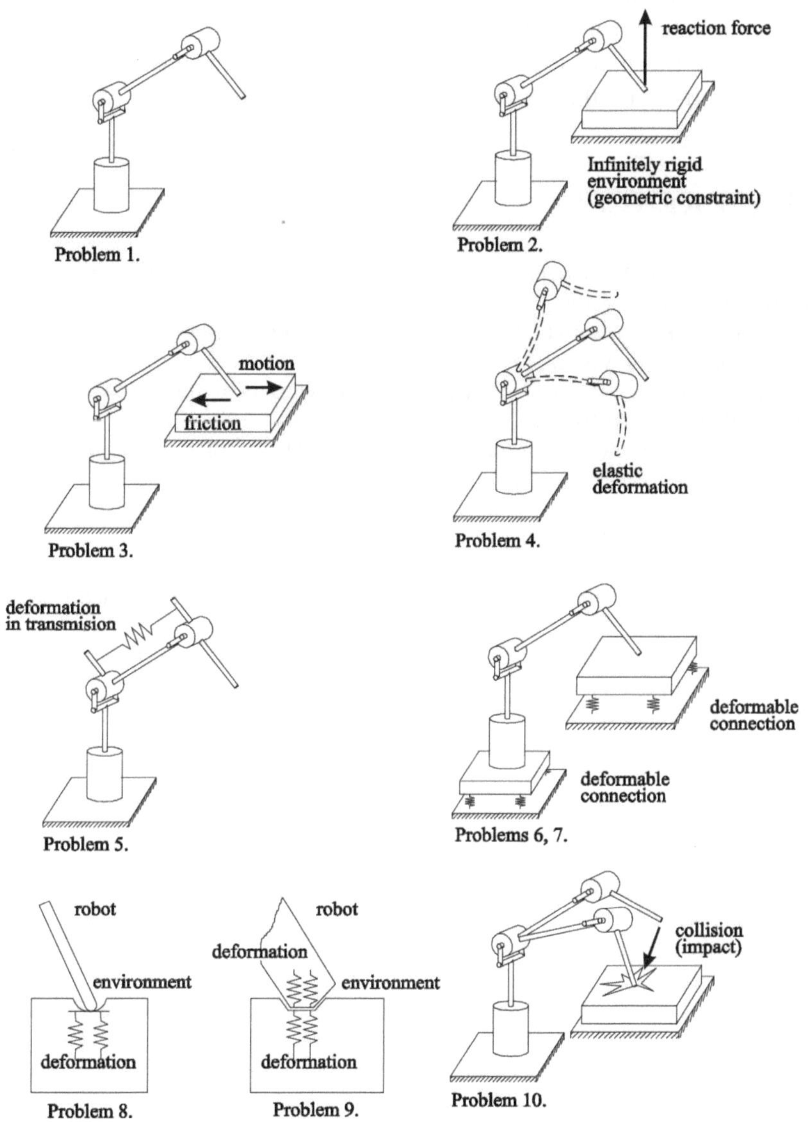

Figure 1.1. Different effects in robot dynamics

The final problem to be mentioned in this survey is redundancy. In the early research in this field, redundancy was considered as a problem of kinematics (avoiding obstacles, avoiding singular positions, etc.). Later research, however, saw redundancy as a possibility of improving robot dynamic performance [55, 56]. Biomechanical approach to solving redundancy of a humanoid robot arm was proposed in [57].

Among various problems in robot dynamics, for this book we selected the following topics: motion subject to geometric constraints and interaction with dynamic environment. Some other problems will be discussed in connection with the main topics (for instance, effects of elastic transmissions, links flexibility, and the problem of kinematic redundancy).

REFERENCES

1. M. Vukobratovic, D. Juricic, "Contribution to the Synthesis of Biped Gait", IEEE Trans. on Biomedical Engineering, Vol. 16, No. 1, 1969.
2. Yu. Stepanenko, "Method of Analysis of Spatial Lever Mechanisms", (in Russian), Mekhanika Mashin, Vol. 23, Moscow, 1970.
3. D. Juricic, M. Vukobratovic, "Mathematical Modeling of a Bipedal Walking System", ASME Publ. 72-WA/BHF-13.
4. M. Vukobratovic, Yu. Stepanenko, "Mathematical Models of General Anthropomorphic Systems", Math. Biosciences, Vol. 17, pp. 191-242, 1973.
5. Yu. Stepanenko, M. Vukobratovic, "Dynamics of Articulated Open-Chain Active Mechanisms", Math. Biosciences, Vol. 28, No. 1/2, 1976.
6. J. Y. S. Luh, M. W. Walker, R. P. C. Raul, "On-Line Computational Scheme for Mechanical Manipulators", Journal of Dynamic Systems, Measurement and Control, Vol. 102, 1980.
7. M. Vukobratovic, V. Potkonjak, "Contribution to Computer Construction of Active Chain Models Via Lagrangian Form", Journal of Applied Mechanics – Trans. of the ASME, No.1, Vol. 46, pp. 181-185, 1979.
8. J. M. Hollerbach, "A Recursive Formulation of Lagrangian Manipulator Dynamics", Proc. of JACC, San Francisco, 1980.
9. E. P. Popov, A. F. Vereschagin, S. A. Zenkevich, Manipulation Robots: Dynamics and Algorithms, (in Russian), "Nauka", Moscow, 1978.
10. V. Potkonjak, M. Vukobratovic, "Two New Methods for Computer Forming of Dynamic Equations of Active Mechanisms", Journal of Mechanisms and Machine Theory, Vol.14, No.3, pp. 189-200, 1979.
11. Robotics Toolbox for MATLAB.
12. Robotica – robot simulation package for MATHEMATICA.
13. M. Vukobratovic, V. Potkonjak, D. Hristic, "Dynamic Method for the Evaluation and Choice of Industrial Manipulators", Proc. 9th ISIR, Washington, 1979.
14. M. Vukobratovic, V. Potkonjak, Dynamics of Manipulation Robots: Theory and Application, Springer-Verlag, 1982.
15. M. Vukobratovic, D. Stokic, Control of Manipulation Robots: Theory and Application, Springer-Verlag, 1982.

16. M. Vukobratovic, V. Potkonjak, Applied Dynamics and CAD of Manipulation Robots, Springer-Verlag, 1985
17. V. Potkonjak, M. Vukobratovic, "Dynamics of Manipulation Mechanisms with Constrained Gripper Motion"–PART1, Journal of Robotic Systems (John Wiley), 3(3), pp. 321-334, and M. Vukobratovic, V. Potkonjak, "Constrained Gripper Motion in Assembly Manipulation"–PART2, Journal of Robotic Systems, 3(3), pp. 335-347, 1986.
18. R. H. Cannon, E. Schmitz, "Initial Experiments on the End-Point Control of a Flexible One-Link Robot", Journal of Robotics Research 3, pp. 62-75, 1984.
19. T. Fukuda, A. Arakawa, "Modeling and Control Characteristics for a Two-Degrees-of-Freedom Coupling System of Flexible Robotic Arm", JSME 30, pp. 1458-1464, 1987.
20. A. Truckenbrodt, "Dynamics and Control Methods for Moving Flexible Structures and Their Application to Industrial Robots", Proc of 5th World Congress on Theory of Machines and Mechanisms, publ. ASME, 1981.
21. H. W. Sunada, Dynamic Analysis of Flexible Spatial Mechanisms and Robotic Manipulators, Ph. D. Thesis, Univ. of California, Los Angeles, 1981.
22. W. J. Book, "Recursive Lagrangian Dynamics of Flexible Manipulator Arms", Journal of Robotics Research, 3, pp. 87-101, 1984.
23. J. O. King, V. G. Gourishankar, R. E. Rink, "Lagrangian Dynamics of Flexible Manipulators Using Angular Velocities Instead of Transformation Matrices", IEEE Trans. on SMC, 17, pp. 1059-1068, 1987.
24. M. Uchiyama, A. Conno, "Computed Acceleration Control for the Vibration Suppression of Flexible Robotic Manipulators", Proc. 5th ICAR, Pisa, Italy, pp. 126-131, 1991.
25. D. Surdilovic, M. Vukobratovic, "One Method for Efficient Dynamic Modeling of Flexible Manipulators", Mechanisms and Machine Theory, Vol. 31, No. 3, pp. 397-315, 1996.
26. D. Surdilovic, M. Vukobratovic, "Deflection Compensation for Large Flexible Manipulators", Mechanisms and Machine Theory, Vol. 31, No. 3, pp. 317-329, 1996.
27. K. Inoue, K. Shiina, M. Takano, K. Sasaki, "Study on Total Computer-aided Design System for Robot Manipulators", Proc. 24th ISIR, pp. 729-736, 1993.
28. M. W. Spong, "Modeling and Control of Elastic Joint Robots", Journal of Dynamic Systems, Measurements and Control, Vol. 109, pp. 309-319, 1987.
29. V. Potkonjak, "Contribution to the Dynamics and Control of Robots Having Elastic Transmission", Robotica, Vol. 6, pp. 63-69, 1988.
30. N. Kircanski, A. Timcenko, M. Vukobratovic, "Position Control of Robot Manipulators with Elastic Joints Using Force Feedback", Journal of Robotic Systems, Vol. 7, No. 4, pp. 535-554, 1990.
31. H. Krishnan, N. H. McClamroch, "A New Approach to Position and Contact Force Regulation in Constrained Robot System", Proc. IEEE Intl. Conf. on Robotics and Automation, Cincinatti, Ohio, pp. 1344-1349, 1990.
32. S. Ahmad, "Constrained Motion (Force/Position) Control of Flexible Joint Robots", Proc. IEEE Conf. on Decision and Control, Brighton, England, pp. 1397-1402, 1991.
33. J. K. Mills, "Stability and Control of Elastic-Joint Robotic Manipulators During Constrained Motion Tasks", IEEE Trans. on Robotics and Automation, Vol. 8, No. 1, pp. 119-125, 1992.
34. B. Brogliato, R. Lozano-Leal, "Adaptive Force/Position Control of Constrained Manipulators with Flexible Joints", Proc. of American Control Conference, Boston, pp. 968-969, 1992.
35. H. Krishnan, "Robot Force and Position Control Including Effects of Joint Flexibility and Actuator Dynamics", Proc. 26th ISIR, Singapore, pp. 185-190, 1995.

36. M. Vukobratovic, V. Potkonjak, V. Matijevic, "Control of Robots with Elastic Joints Interacting with Dynamic Environment", Journal of Intelligent & Robotic Systems, Vol. 23, No. 1, pp. 87-100, 1998.
37. M. H. Raibert, J. J. Craig, "Hybrid Position/Force Control of Manipulators", Journal of Dynamic Systems, Measurements and Control, Vol. 103, pp. 126-133, 1981.
38. M. T. Mason, "Compliance and Force Control for Computer Controlled Manipulators", IEEE Trans. on Systems, Man and Cybernetics, Vol. 11, No. 6, pp. 418-432, 1981.
39. T. Yoshikava, T. Sugie, M. Tanaka, "Dynamic Hybrid Position/Force Control of Robot Manipulators – Controller Design and Experiment", IEEE Journal of Robotics and Automation, Vol. 4, No. 6, pp. 699-705, 1998.
40. A. De Luca, C. Manes, Hybrid Force/Position Control for Robots in Contact with Dynamic Environments, Proc. of Robot Control, SYROCO'91, Vienna, pp. 377-382, 1991.
41. N. Hogan, "Impedance Control: An Approach To Manipulation", Part I–Part III, Journal of Dynamic Systems, Measurements and Control, Vol. 107, pp. 1-24, 1985.
42. H. Kazeroni, T.B. Sheridan, P.K. Houpt, "Robust Compliant Motion for Manipulators", Part 1: The Fundamental Concepts of Compliant Motion, IEEE Journal of Robotics and Automation, RA 2, No. 2, pp. 83-92, 1986.
43. Y. Ekalo, M. Vukobratovic, "Adaptive Stabilization of Motion and Force in Contact Tasks for Robotic Manipulators with Non-Stationary Dynamics", Intl. J. of Robotics and Automation, Vol. 9, No. 3, pp. 91-98, 1994.
44. M. Vukobratovic, Y. Ekalo, "New Approach to Control of Robotic Manipulators Interacting with Dynamic Environment", Robotica, Vol. 14, No. 1, pp. 31-39, 1996.
45. M. Vukobratovic, R. Stojic, Y. Ekalo, "Contribution to the Problem Solution of Position/Force Control of Manipulation Robots in Contact with Dynamic Environment – A Generalization", IFAC Automatica, Vol. 34, No. 10, 1998.
46. V. N. Chumenko, A. S. Yuschenko, "Impact Effects Upon Manipulation Robot Mechanism" (in Russian), Technical Cybernetics, No. 4, 1981.
47. Y. F. Zheng, H. Hemami, "Mathematical Modeling of a Robot Collision with its Environment", J. of Robotic Systems, 2(3), pp. 289-307, 1985.
48. J. B. Keller, "Impact with Friction", Journal of Applied Mechanics, Vol. 53/1–53/4, 1986.
49. W. J. Stronge, "Rigid Body Collision with Friction", Proc. R. Soc. Lond. A., Vol. 431, pp. 169-181, 1990.
50. Y. Hurmuzlu, D. B. Marghitu, "Rigid Body Collision of Planar Kinematic Chain with Multiple Contact Points", Intl. J. of Robotic Research, Vol. 13, No. 1, pp. 82-92, 1994.
51. V. Drenovac, V. Potkonjak, "Contribution to the Modeling of Impact with Robotic Systems", Robotica, Vol. 11, Part. 4, pp. 445-452, 1993.
52. B. Brogliato, P. Orhant, "On the Transition Phase in Robotics: Impact Models, Dynamics and Control", Proc. IEEE Conf. on Robotics and Automation, San Diego, 1994.
53. G. M. Acaccia, P. C. Gagetti, M. Callegari, R. C. Michelini, R. M. Molfino, "Modeling the Impact Dynamics of Robotic Manipulators", Proc. 4th IFAC Symp. on Robot Control, Capri, Italy, pp. 559-564, 1995.
54. A Tornambe, "Modeling and Controlling Two-Degrees-of-Freedom Impacts", Proc. 3rd IEEE Mediterranean Conf. on Control and Automation, Lymassol, Cyprus, 1995.
55. N. Hogan, A. Sharon, E. D. Hardt, "High Bandwidth Force Regulation and Inertia Reduction Using a Macro/Micro Manipulator System", Proc. IEEE Conf. on Robotics and Automation, pp. 126-132, 1988.

56. V. Potkonjak, A. Krstulovic, "Contribution to the Kinematics and Dynamics of Redundant Robots Via Distributed Positioning", Journal of Intelligent and Robotic Systems 5, pp. 229-239, 1992.
57. V. Potkonjak, M. Popovic, M. Lazarevic, J. Sinanovic, "Redundancy Problem in Writing: From Human to Anthropomorphic Robot Arm", IEEE Transactions on SMC, Part B: Cybernetics, Vol. 28, No. 6, pp. 790-805, Dec. 1998.

Chapter 2

FREE MOTION OF A RIGID–BODY ROBOT ARM

This chapter considers the problems of dynamics assuming that robot links are infinitely rigid. The discussion starts with relevant ideas from the theory of mechanisms (kinematic pairs and chains, degrees of freedom, etc.) and then defines the parameters that describe the robot link geometry. Kinematics and dynamics concern free motion of the robot arm. Dynamic model of the robot chain is expanded by introducing the actuator dynamics and elastodynamics of the transmission system. Finally, some issues of control are elaborated.

1. ROBOT GEOMETRY AND KINEMATICS

1.1 Degrees of Freedom and the Robot Arm

The first notion to be defined is kinematic pair. It is formed of two bodies connected in such a way that one restricts the motion of the other. Depending on the other characteristics of the relative motion, kinematic pairs are separated into classes. The connection between bodies may restrict relative motion to a larger or smaller extent. The weakest restriction is the one that does not exist. It is, of course, a hypothetical situation since in such a case the kinematic pair does not exist either – there are two free bodies. However, we shall conditionally call this "connection" zero-class pairs. In such a situation the motion of one body relative to the other may be considered as a combination of the translations along three directions perpen-dicular to each other, and the rotations about these directions (Fig. 2.1). Thus, we talk about six degrees of freedom.

If the connection restricts the relative motion in such a way that five DOFs remain, then we talk about the first-class pair. Generally, if a pair has k DOFs then its class is 6-k. Some examples are given in Fig. 2.2. In robotics, the most important are fifth-class pairs that allow one relative rotation or one relative translation between the two bodies. The first one will be called *revolute joint* and the latter one *linear joint*. All the discussion that follows will be restricted to these two types of joints since fifth-class pairs are the only ones that can be actuated. This is due to the fact that today's actuators themselves are the fifth-class pairs (stator and rotor of an electrical motor, or, cylinder and piston of a hydraulic actuator). Schematic representations of a revolute joint and a linear joint are given in Fig. 2.3.

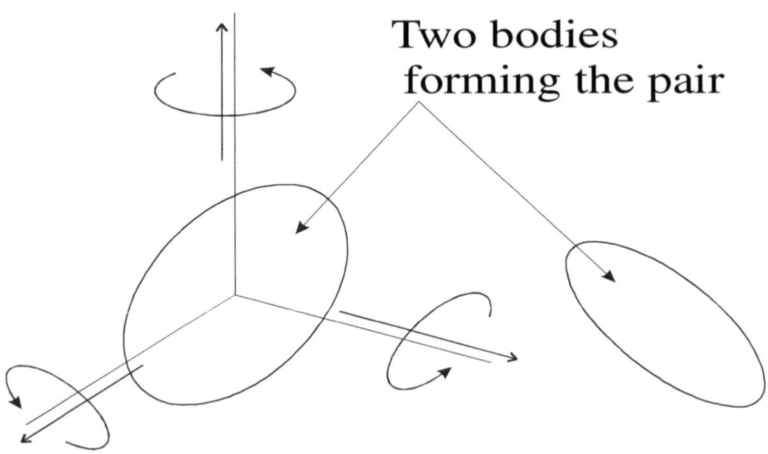

Figure 2.1. Zero-class" pair (two free bodies). One body can move relative to the other in six different and independent ways. This means that the pair has six DOFs.

Kinematic chain (or linkage) is formed of several bodies interconnected by joints. Bodies, members of the chain, are called *links*. If each link is involved in as many as two kinematic pairs, then we talk about *simple chain*. Figure 2.4 shows a simple chain of a six-joint robot. If one link is connected to three neighbors, we talk of a *branched chain*. We shall restrict our consideration to simple chains. The other way of partitioning chains into groups concerns the possibility of forming a closed loop. If it is not possible to find a closed loop, then the chain is *open* – in a contrary case we talk about a *closed chain*. Figure 2.4 demonstrates that a robot chain changes its structure (turning from open structure to closed and back) during practical industrial work, e.g., peg-in-hole tasks. In this chapter we consider open chains only.

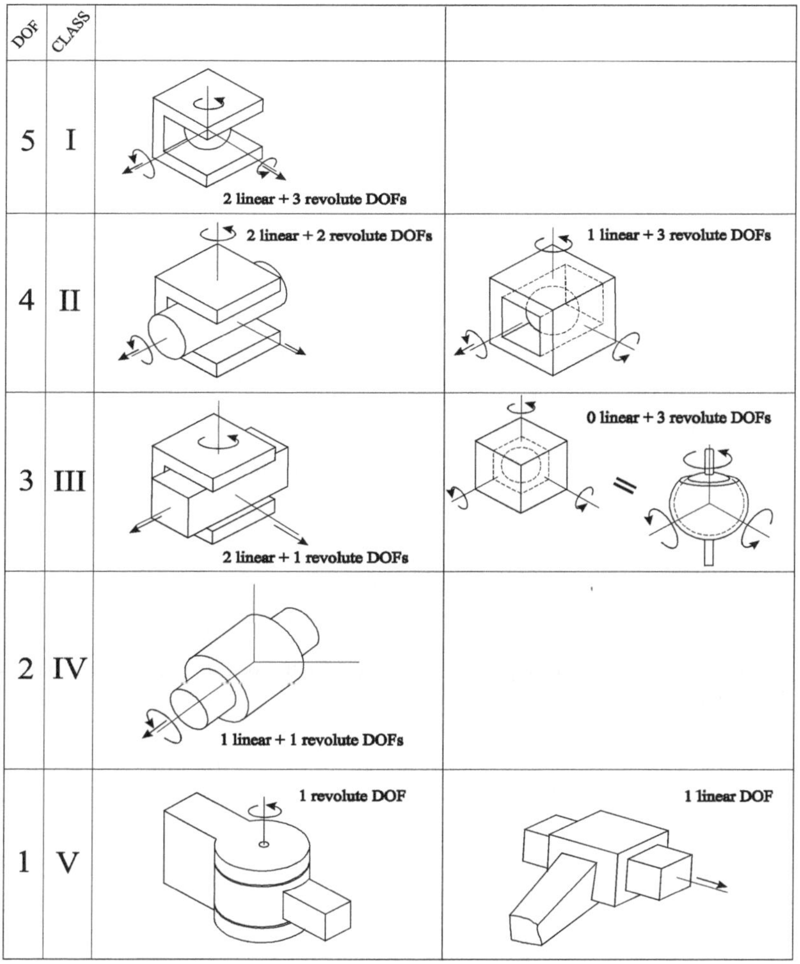

Figure 2.2. Different classes of pairs. Successively restricting the possibilities of motion, from 5 DOFs to 1, the class of the pair is changed from 1 to 5. Robotics is primarily interested in fifth-class pairs.

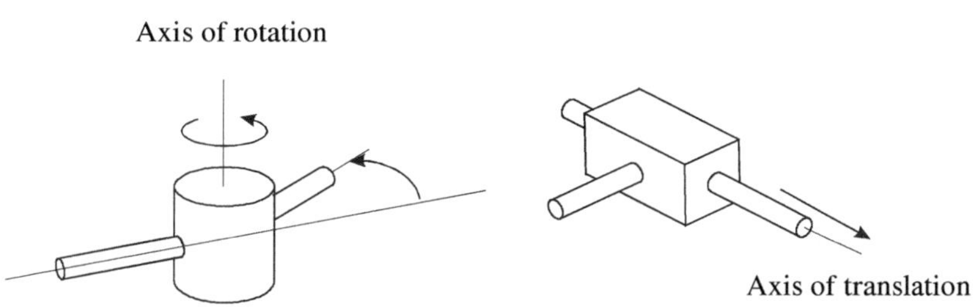

Figure 2.3. Schematic representation of a revolute joint (a) and a linear joint (b).

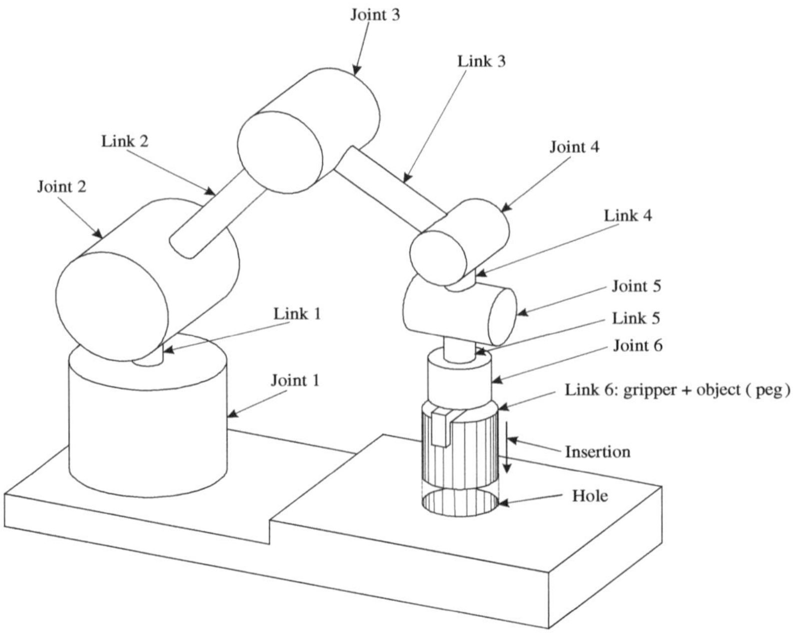

Figure 2.4. A robot chain changes its structure (from open to closed and back) when executing an industrial task

Since the consideration is restricted to open chains and one-DOF joints, one concludes that for the topical mechanical system the number of DOFs equals the number of joints and the number of links (Fig. 2.5). In order to describe the position of the chain, it is possible to introduce one coordinate for each joint. Let us number the links by 1, 2, ... , n, (then the support has number zero). Joints are designated as S_1, S_2, ... , S_n . A coordinate q_j defines the position of the joint S_j. If S_j is a revolute joint then the coordinate q_j represents the angle of rotation, and if it is a linear joint then q_j is the longitudinal displacement. Now, the n-dimensional vector

$$q = [q_1 \; q_2 \; \dots \; q_n]^T \tag{2.1}$$

defines the chain position. These coordinates are called generalized coordinates, but since they define the relative position of the neighboring links, the term *internal coordinates* is often used.

A robot chain may be decomposed into some functional parts. *Minimal configuration* consists of the first three joints and first three links (Fig. 2.6). It is often called *arm*. The joints may be linear or revolute. The *wrist* is attached to the top of the minimal configuration and it involves one to three revolute joints. The *end effector* is the terminal link of the chain. It is a device specific to the operation the robot has to perform (gripping, spray-painting, etc.).

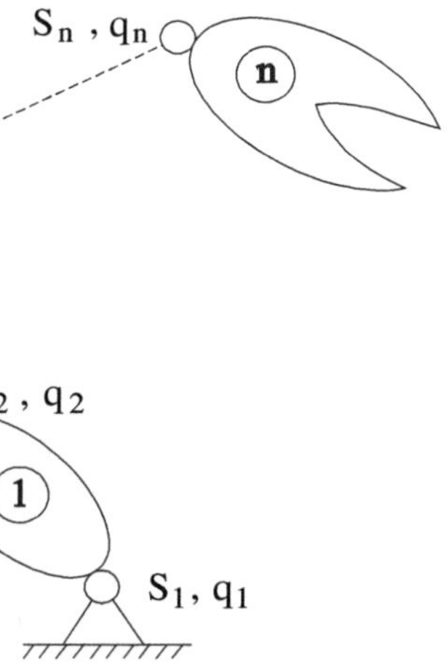

Figure 2.5. If an open chain consists of n links interconnected by means of n one-DOF joints, then it has n DOFs. Its position can be defined by using n coordinates, selecting one at each joint.

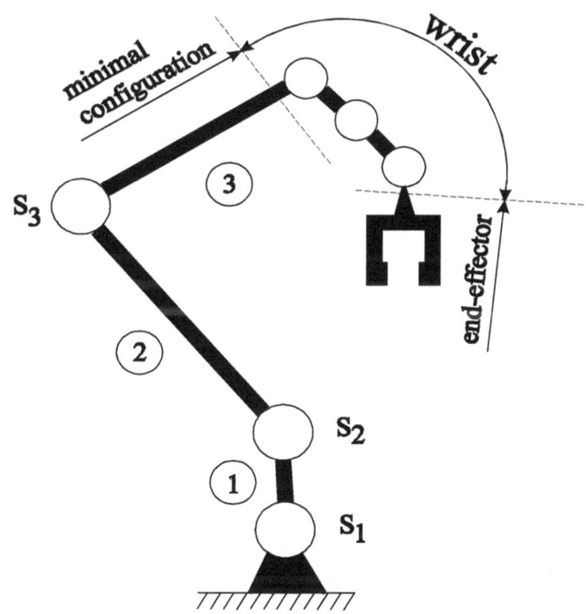

Figure 2.6 Robot chain is decomposed into three functional parts: minimal configuration, wrist and end effector.

Figure 2.7 shows several typical structures of robot minimal configuration and their working spaces.

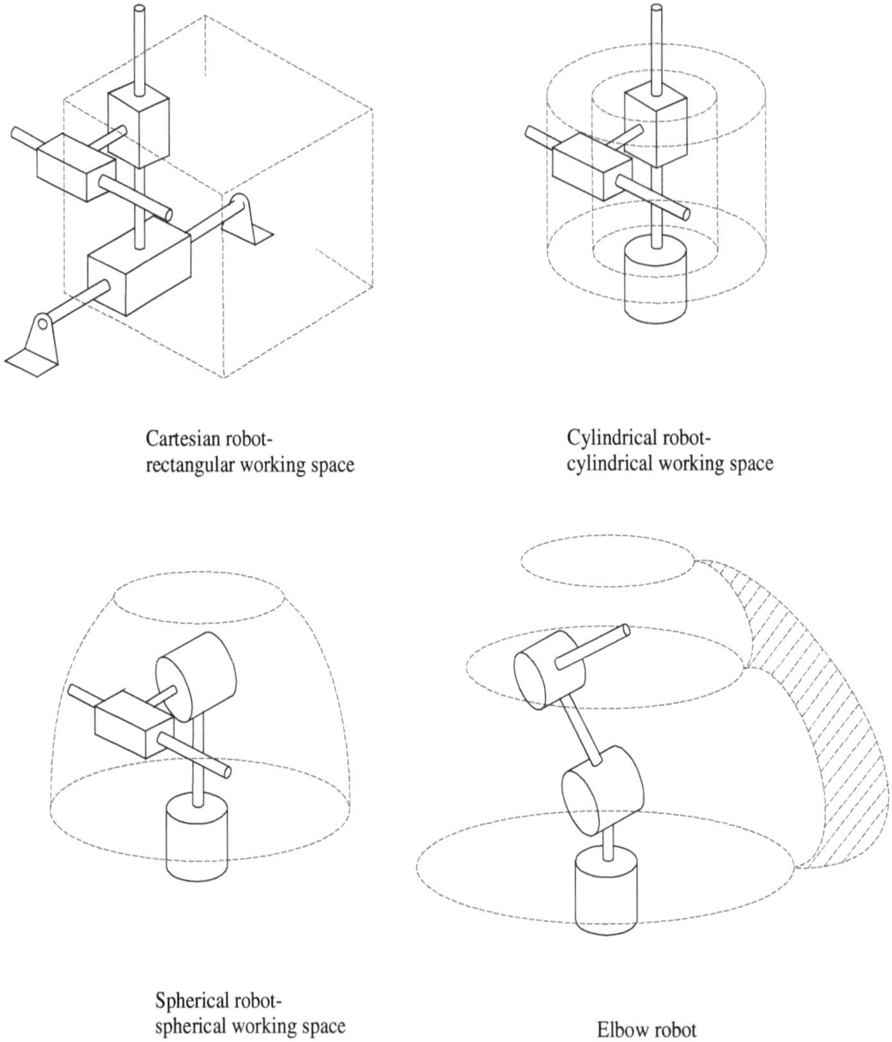

Figure 2.7. Typical structures of robot minimal configurations and their working areas

1.2 Geometry

In order to define the geometry of a robot chain consisting of *n* links, it is necessary to define the geometry of each link in the sense of relative position and direction of the joints. Two approaches have been proposed to the definition of this geometry. The first one (introduced in robotics in [1, 2]) is based on Rodrigue's formulae of finite rotations. The other one uses the

Denavit–Hartenberg transform [3]. Although the first approach is more general (especially in further elaboration of dynamics), the latter is more widely accepted. We shall describe both approaches.

(a) Method based on Rodrigue's formulae of finite rotations

Consider the j-th link of the chain (Fig. 2.8). It connects two joints, S_j and S_{j+1}. The positions of the two joints are defined by means of the vectors $\vec{r}_{j,j}$ and $\vec{r}_{j,j+1}$. Directions are defined by unit vectors of the joint axes $\vec{e}_j$ and $\vec{e}_{j+1}$. These vectors move with the body and thus they are not constant. However, if considered relative to the body they become constant. For this reason we introduce a Cartesian frame that is fixed to the link and has the origin at the mass center (MC, point C_j in Fig. 2.8). The frame axes may be oriented arbitrarily (keeping them perpendicular to each other). We now introduce the notation for a vector expressed in the link-fixed frame. If a vector has index j then the notation $\vec{d}_j$ will be used to express three projections onto the axes of an external fixed frame, while the notation $\vec{\vec{d}}_j$ will indicate the projections onto the frame fixed to link j. When needed, the projections onto the preceding frame (j-1) will be denoted as $\underline{\vec{d}}_j$. This convention holds for all the vectors appearing in the analysis of geometry, kinematics, and dynamics (e.g., vectors of joint axes, velocities, forces, etc.). Now, the four vectors, $\vec{\vec{e}}_j$, $\underline{\vec{e}}_{j+1}$, $\vec{\vec{r}}_{j,j}$, and $\vec{\vec{r}}_{j,j+1}$ define the geometry of the link j. By defining all links j=1, ... , n , the complete chain is defined.

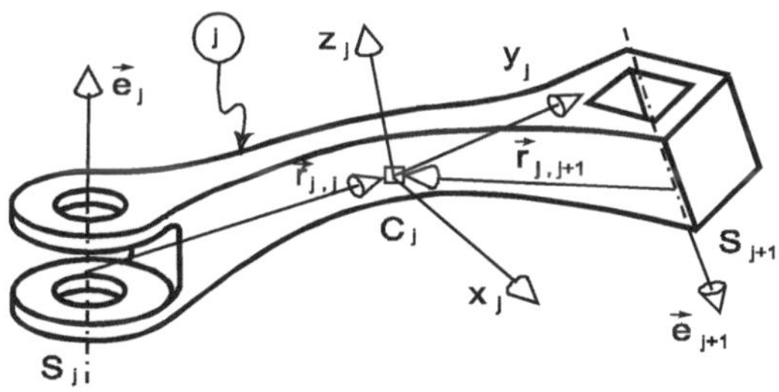

Figure 2.8. Four vectors define the geometry of a link

Now, it is possible to give the exact definition of joint coordinates. First, we recall that the discussion is restricted to revolute and linear one-DOF joints. Indicators s_j, j=1, ... , n are used to define the type of joints:

- s_j=0 means that the joint S_j is revolute, and
- s_j=1 indicates a linear joint.

For a revolute joint S_j, the coordinate q_j (angle of rotation) is defined as shown in Fig. 2.9. The zero position corresponds to the "extended" joint. However, if it happens that any of the vectors $\vec{r}_{j-1,j}$ and $\vec{r}_{j,j}$ is parallel to the joint axis $\vec{e}_j$, the definition does not apply. If $\vec{r}_{j-1,j} \| \vec{e}_j$, we talk about the "specificity" of the link j-1 on its upper side, and introduce an additional vector $\vec{r}^{\,*}_{j-1,j}$ to use it instead of $\vec{r}_{j-1,j}$ when defining the coordinate. The only restriction in the choice of the additional vector is that it must not be parallel to the joint axis. We usually define it to be perpendicular to $\vec{e}_j$. If $\vec{r}_{j,j} \| \vec{e}_j$, we talk about the "specificity" of the link j on its lower side, and introduce $\vec{r}^{\,*}_{j,j}$ to replace $\vec{r}_{j,j}$ in the definition of the coordinate. $\vec{r}^{\,*}_{j,j}$ must not be parallel to the joint axis.

For a linear joint S_j, the coordinate q_j is defined as a displacement along the joint axis, as shown in Fig. 2.10. Zero position may be chosen arbitrarily.

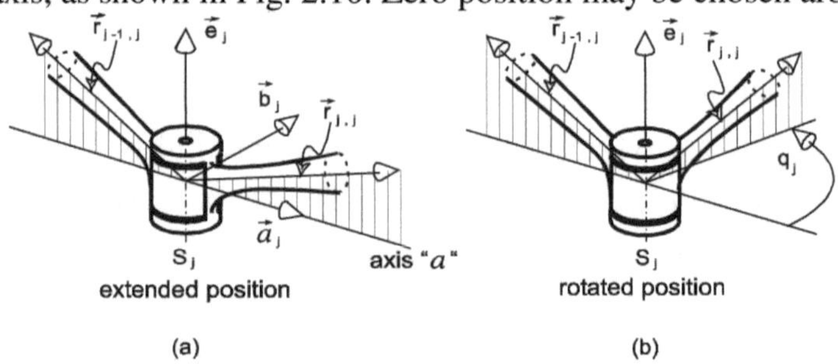

Figure 2.9. Definition of the angular coordinate at a revolute joint

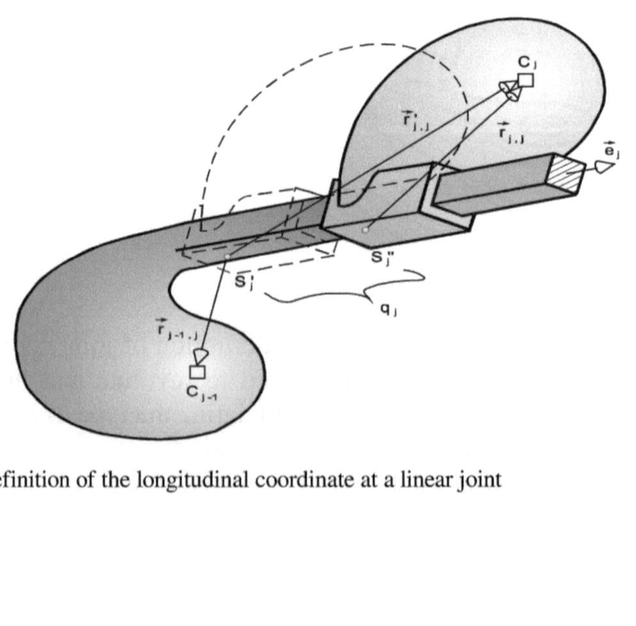

Figure 2.10. Definition of the longitudinal coordinate at a linear joint

After introducing the link-fixed frames, a question naturally arises about how to transform a vector from one frame to another. To transform a vector $\vec{d}_j$ from the frame j to the frame j-1, the transformation matrix $A_{j-1,j}$ is applied:

$$\underline{\vec{d}}_j = A_{j-1,j}\vec{\vec{d}}_j \tag{2.2}$$

If the vector should be transferred from the frame j to the external immobile frame then:

$$\vec{d}_j = A_j\vec{\vec{d}}_j \tag{2.3}$$

and it is easy to prove that

$$A_j = A_{0,j} = A_{0,1}A_{1,2}...A_{j-1,j}\ . \tag{2.4}$$

Since the transformation matrices are orthogonal, the inverse transformation is made by using their transposes, $A^{-1} = A^T$.

In order to calculate the transformation matrix $A_{j-1,j}$, we need the triple of linearly independent vectors with projections known in both the frames j-1 and j. First, we do the calculation for the extended joint, that is, for q_j=0 (Fig. 2.9a). The first vector of the triple is $\vec{e}_j$ since its projections, $\underline{\vec{e}}_j$ to the frame j-1 and $\vec{\vec{e}}_j$ to the frame j, are known (when the geometry is defined). We now introduce a vector $\vec{a}_j$ as a unit vector of the axis "a" shown in Fig. 2.9a. Its projections are:

$$\underline{\vec{a}}_j = \frac{-\underline{\vec{e}}_j \times (\vec{\vec{r}}_{j-1,j} \times \underline{\vec{e}}_j)}{\left|\underline{\vec{e}}_j \times (\vec{\vec{r}}_{j-1,j} \times \underline{\vec{e}}_j)\right|}, \quad \vec{\vec{a}}_j = \frac{\vec{\vec{e}}_j \times (\vec{\vec{r}}_{j,j} \times \vec{\vec{e}}_j)}{\left|\vec{\vec{e}}_j \times (\vec{\vec{r}}_{j,j} \times \vec{\vec{e}}_j)\right|} \tag{2.5}$$

The third vector is obtained by means of the vector product, $\vec{b}_j = \vec{e}_j \times \vec{a}_j$. Its projections are:

$$\underline{\vec{b}}_j = \underline{\vec{e}}_j \times \underline{\vec{a}}_j\ , \quad \vec{\vec{b}}_j = \vec{\vec{e}}_j \times \vec{\vec{a}}_j\ . \tag{2.6}$$

Vectors $\vec{e}_j$, $\vec{a}_j$ and $\vec{b}_j$ are perpendicular to each other and hence linearly independent. According to (2.2), the relation between the projections of the triple may be expressed as

$$\vec{\underline{e}}_j = A^0_{j-1,j}\vec{\bar{e}}_j\,,\quad \vec{\underline{a}}_j = A^0_{j-1,j}\vec{\bar{a}}_j\,,\quad \vec{\underline{b}}_j = A^0_{j-1,j}\vec{\bar{b}}_j\,, \tag{2.7}$$

where the superscript "0" with the transformation matrix $A^0_{j-1,j}$ indicates the extended position of the joint. If matrix calculation is to be used, it is appropriate to introduce some notation for the 3×1 matrix that replaces a vector. This is done by omitting the symbol "→" above the vector. Thus, d represents a 3×1 matrix that corresponds to the vector $\vec{d}$. This convention will hold for all vectors.

Now, (2.7) may be rewritten in a matrix form

$$[\underline{e}_j\ \underline{a}_j\ \underline{b}_j] = A^0_{j-1,j}[\bar{e}_j\ \bar{a}_j\ \bar{b}_j], \tag{2.8}$$

whereby the transformation matrix can be obtained as

$$A^0_{j-1,j} = [\underline{e}_j\ \underline{a}_j\ \underline{b}_j][\bar{e}_j\ \bar{a}_j\ \bar{b}_j]^{-1}. \tag{2.9}$$

The matrix $A^0_{j-1,j}$ consists of three columns. They represent unit vectors of the frame j expressed with respect to the frame j-1. Let us denote these columns by $\underline{V}^0_{j1}$, $\underline{V}^0_{j2}$ and $\underline{V}^0_{j3}$, that is

$$A^0_{j-1,j} = [\underline{V}^0_{j1}\ \underline{V}^0_{j2}\ \underline{V}^0_{j3}]. \tag{2.10}$$

We now rotate the joint by the angle q_j (Fig. 2.9b) and search for the change in the transformation matrix. Let the new matrix be $A_{j-1,j}$. In order to calculate the new matrix, it is necessary to rotate each of the unit vectors $\vec{\underline{V}}^0_{jk}$, k=1,2,3 for the angle q_j about the axis $\vec{e}_j$. This is done by applying Rodrigue's formulae

$$\begin{aligned}\vec{\underline{V}}_{jk} = \vec{\underline{V}}^0_{jk}\cos q_j + (1-\cos q_j)(\vec{\underline{e}}_j \times \vec{\underline{V}}^0_{jk})\cdot\vec{\underline{e}}_j + \\ + \vec{\underline{e}}_j \times \vec{\underline{V}}^0_{jk}\sin q_j, \quad k = 1,2,3,\end{aligned} \tag{2.11}$$

where $\vec{\underline{V}}_{jk}$ represents the unit vector after rotation, and thus, the k-th column of the matrix $A_{j-1,j}$. Transformation matrix is now

$$A_{j-1,j} = [\underline{V}_{j1}\ \underline{V}_{j2}\ \underline{V}_{j3}]. \tag{2.12}$$

Since the matrix depends on the coordinate q_j, one may write $A_{j-1,j} = A_{j-1,j}(q_j)$. From (2.4), it follows that $A_j = A_j(q_1,...,q_j) = A_j(q)$.

In the case of "specificity", the vector $\vec{r}^{\,*}_{j-1,j}$ is used instead of $\vec{r}_{j-1,j}$, and $\vec{r}^{\,*}_{j,j}$ instead of $\vec{r}_{j,j}$.

Consider now a linear joint S_j (Fig. 2.10). It is clear that the transformation matrix between the frames j and j-1 is constant, that is, it does not depend on the longitudinal coordinate q_j. Calculation of the matrix can be performed in the same manner as for the revolute joint.

However, one can see from Figure 2.11 that the rotation about the axis $\vec{e}_j$ is performed for an angle α_j, which is constant and has to be given (Fig. 2.11a). Next, one finds that the "specificity" is a very common feature of linear joints (Fig. 2.11b).

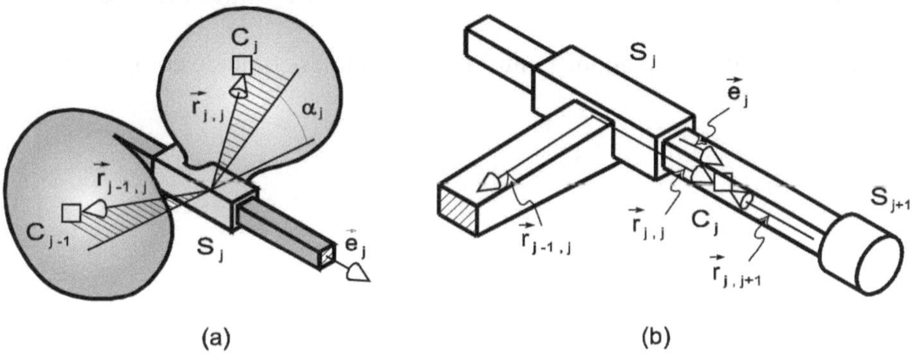

Figure 2.11. Some features of linear joints

(b) Method based on the Denavit–Hartenberg transform

In this approach, numbering of joints usually starts from zero, that is, S_0, S_1, ... Thus, for a link j, the joint on the lower side is S_{j-1}, and on the upper side S_j (Fig. 2.12). Note that the coordinates are still counted from 1. This means that the coordinate q_j corresponds to the joint S_{j-1}.

In order to express the vectors of geometry, we introduce the link-fixed frames. For a link j the origin of the frame is placed at the point S_j (Fig. 2.12). The axes of the frame cannot be oriented arbitrarily but their orientation defines some geometrical features of the link and the corresponding joints. The axis z_j of the link-fixed frame is oriented along the joint axis $\vec{e}_j$. The axis x_j is placed so as to be perpendicular to both $\vec{e}_j$ and $\vec{e}_{j-1}$. Finally, y_j is placed so as to obtain a Cartesian frame (that is, perpendicular to z_j and x_j).

Geometry of the link j is defined by means of the vectors $\vec{r}_j = \overrightarrow{S_j C_j}$ and $\vec{\ell}_j = \overrightarrow{S_{j-1} S_j}$. If these vectors are expressed in the j-th link-fixed frame, the vector constants $\tilde{\vec{r}}_j$ and $\tilde{\vec{\ell}}_j$ are obtained to define the link geometry. In this way the link CM and relative position of the two neighboring joints are defined. Directions of the joint axes are determined by means of the frame axes.

Let us consider a revolute joint S_j and the transformation of the frame $O_{j-1}x_{j-1}y_{j-1}z_{j-1}$ to the frame $O_j x_j y_j z_j$. Four phases of the transformation can be distinguished (Fig. 2.13):

- phase (a): Rotation about z_{j-1} until x_{j-1} becomes parallel to x_j. Angle of rotation is q_j.
- phase (b): Translation along z_{j-1} until x_{j-1} coincides with x_j. Length of translation is c_j.
- phase (c): Translation along x_j until O_{j-1} coincides with O_j. Length of translation is a_j.
- phase (d): Rotation about x_j until the two frames coincide completely. Angle is β_j.

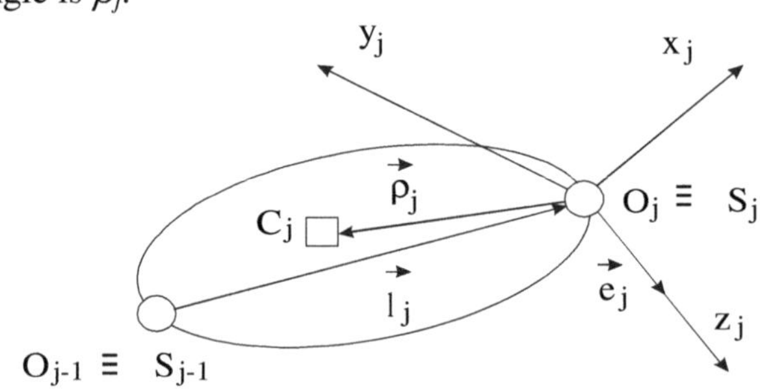

Figure 2.12. Link-fixed frame in the Denavit–Hartenberg approach

The transformation matrix that corresponds to phase (a) is

$$A_{j,j-1}^{(a)} = \begin{bmatrix} \cos q_j & \sin q_j & 0 \\ -\sin q_j & \cos q_j & 0 \\ 0 & 0 & 1 \end{bmatrix} \tag{2.13}$$

and the transformation matrix for phase (d) is

$$A_{j,j-1}^{(d)} = \begin{bmatrix} 1 & 0 & 0 \\ 0 & \cos\beta_j & \sin\beta_j \\ 0 & -\sin\beta_j & \cos\beta_j \end{bmatrix}. \tag{2.14}$$

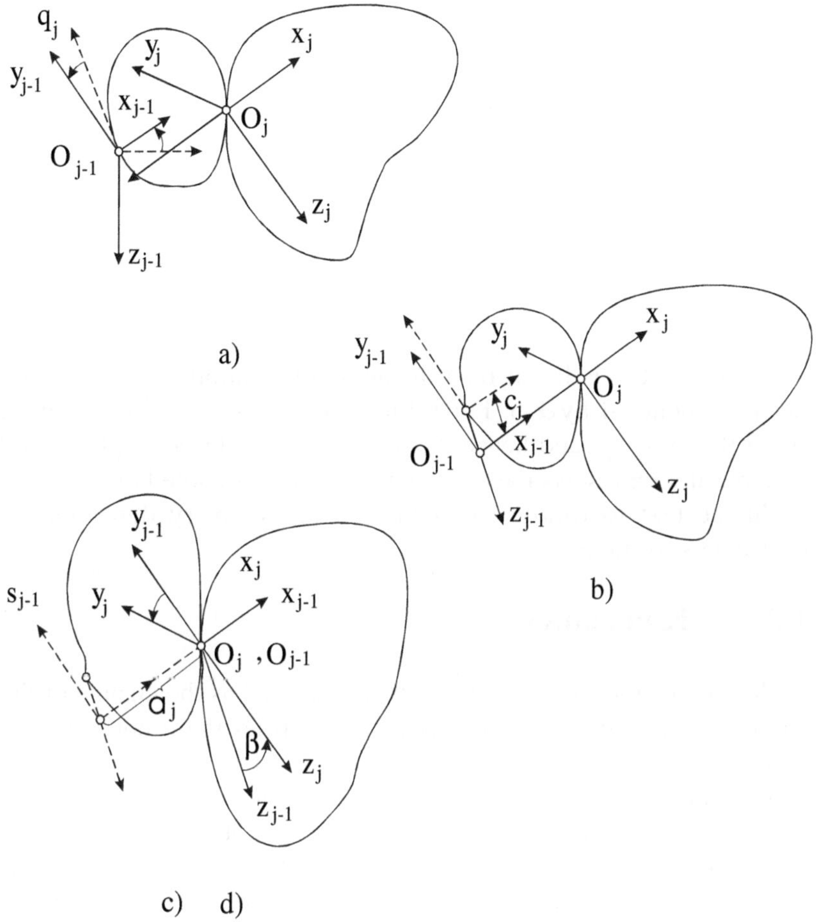

Figure 2.13. Four phases of transformation of the frame j-1 to the frame j

Phases (b) and (c) do not contribute to the transformation matrix due to their translational character. The complete transformation matrix is

$$A_{j,j-1} = A_{j,j-1}^{(d)} A_{j,j-1}^{(a)} , \tag{2.15}$$

that is

$$A_{j,j-1} = \begin{bmatrix} \cos q_j & \sin q_j & 0 \\ -\sin q_j \cos\beta_j & \cos q_j \cos\beta_j & \sin\beta_j \\ \sin q_j \sin\beta_j & -\cos q_j \sin\beta_j & \cos\beta_j \end{bmatrix}. \tag{2.16}$$

The inverse transformation is

$$A_{j-1,j} = A_{j,j-1}^{-1} = A_{j,j-1}^{T} \tag{2.17}$$

Phases (b) and (c), considered together, represent the translation of the frame origin by the vector $\vec{\ell}_j$, which can be expressed as

$$\underline{\ell}_j = \begin{bmatrix} a_j \cos q_j & a_j \sin q_j & c_j \end{bmatrix}^T . \tag{2.18}$$

If the joint S_j is linear then the angle of rotation in phase (a) is constant (let it be denoted by α_j). For a linear joint, the joint motion is performed along the axis z_{j-1} and hence the length c_j introduced in phase (b) becomes variable and represents the coordinate. It is then denoted by q_j.

In the text to follow we shall use the geometry description based on Rodrigue's formulae.

1.3 Kinematics

Motion of the robot chain can be considered as the change of the internal coordinates q. Thus, time history $q=q(t)$ defines the robot motion. The first derivative of the position vector, that is $\dot{q} = [\dot{q}_1 \dots \dot{q}_n]^T$, defines the speed, and is called the vector of generalized or internal velocities. The second derivative $\ddot{q}$ is the vector of internal accelerations. However, it is often interesting to know the motion of the robot links, i.e. the velocity and acceleration of the MC, and angular velocity and acceleration for each link. These variables are calculated in a recursive manner. Supposing that the kinematic variables are known for the link j-1 and we want to find the values for the link j. Let $\vec{r}_{j-1}^{\,C}$ be the position vector of the MC $C_{j\text{-}1}$, $\vec{v}_{j-1}$ be its velocity, and $\vec{w}_{j-1}$ the acceleration. Further, let $\vec{\omega}_{j-1}$ be the angular velocity of the link j-1 and $\vec{\varepsilon}_{j-1}$ the angular acceleration. Recursive expressions that give the kinematic values for the link j are [1, 2, 4]:

$$\vec{\omega}_j = \vec{\omega}_{j-1} + \dot{q}_j (1 - s_j) \vec{e}_j , \tag{2.19}$$

$$\vec{\varepsilon}_j = \vec{\varepsilon}_{j-1} + (\ddot{q}_j \vec{e}_j + \dot{q}_j (\vec{\omega}_j \times \vec{e}_j))(1 - s_j) , \tag{2.20}$$

$$\vec{r}_j^{\,C} = \vec{r}_{j-1}^{\,C} - \vec{r}_{j-1,j} + \vec{r}\,'_{j,j} , \tag{2.21}$$

$$\vec{v}_j = \vec{v}_{j-1} - \vec{\omega}_{j-1} \times \vec{r}_{j-1,j} + \vec{\omega}_j \times \vec{r}\,'_{j,j} + \dot{q}_j s_j \vec{e}_j , \tag{2.22}$$

$$\vec{w}_j = \vec{w}_{j-1} - \vec{\varepsilon}_{j-1} \times \vec{r}_{j-1,j} - \vec{\omega}_{j-1} \times (\vec{\omega}_{j-1} \times \vec{r}_{j-1,j}) + \\ + \vec{\varepsilon}_j \times \vec{r}\,'_{j,j} + \vec{\omega}_j \times (\vec{\omega}_j \times \vec{r}\,'_{j,j}) + (\ddot{q}_j \vec{e}_j + 2\dot{q}_j (\vec{\omega}_j \times \vec{e}_j)) s_j \ , \tag{2.23}$$

where

$$\vec{r}\,'_{j,j} = \vec{r}_{j,j} + s_j q_j \vec{e}_j , \tag{2.24}$$

and s_j is the indicator of joint type (0 for revolute and 1 for linear joints).

The initial values for recursive calculation, $\vec{\omega}_0$, $\vec{\varepsilon}_0$, $\vec{r}_0^{\,C}$, $\vec{v}_0$ and $\vec{w}_0$, are usually assumed to be zero. However, there may exist non-zero values and then they refer to a moving support.

Velocities $\vec{\omega}_j$ and $\vec{v}_j$ may be expressed as linear forms with respect to the internal velocities, and accelerations $\vec{\varepsilon}_j$ and $\vec{w}_j$ as linear forms with respect to the internal accelerations:

$$\vec{\omega}_j = \sum_{k=1}^{j} \vec{\lambda}_k^j \dot{q}_k , \tag{2.25}$$

$$\vec{v}_j = \sum_{k=1}^{j} \vec{\xi}_k^{\,j} \dot{q}_k , \tag{2.26}$$

$$\vec{\varepsilon}_j = \sum_{k=1}^{j} \vec{\lambda}_k^j \ddot{q}_k + \vec{\gamma}^{\,j} , \tag{2.27}$$

$$\vec{w}_j = \sum_{k=1}^{j} \vec{\xi}_k^{\,j} \ddot{q}_k + \vec{\delta}^{\,j} . \tag{2.28}$$

We now turn to matrix notation (symbol "→" is omitted if the vector is treated as a 3×1 matrix) and rewrite equations (2.25)–(2.28):

$$\omega_j = \Lambda^j \dot{q} , \tag{2.29}$$

$$v_j = \Xi^j \dot{q} , \tag{2.30}$$

$$\varepsilon_j = \Lambda^j \ddot{q} + \Gamma^j , \tag{2.31}$$

$$w_j = \Xi^j \ddot{q} + \Delta^j , \tag{2.32}$$

where q is the n-vector of internal coordinates. Matrices Λ^j, Γ^j, Ξ^j, and Δ^j are of dimensions $3\times n$, 3×1, $3\times n$, and 3×1, respectively. Their columns are the coefficients and the free terms of the linear forms (2.25)–(2.28):

$$\Lambda^j = \left[\lambda_1^j \ \ldots \ \lambda_j^j \ 0 \ldots 0\right]_{\dim=3\times n}, \tag{2.33}$$

$$\Gamma^j = \left[\gamma^j\right]_{\dim=3\times 1}, \tag{2.34}$$

$$\Xi^j = \left[\xi_1^j \ \ldots \ \xi_j^j \ 0 \ldots 0\right]_{\dim=3\times n}, \tag{2.35}$$

$$\Delta^j = \left[\delta^j\right]_{\dim=3\times 1}. \tag{2.36}$$

In the iterative calculation of these matrices, the superscript j is just the iteration number and hence the matrices are often written without it: Λ, Γ, Ξ, and Δ. Starting from the recursive expressions (2.19)–(2.23) we find that in the j-th iteration (passing from j-1 to j) it is necessary to make the following changes and supplements of the four matrices:

$$\begin{aligned}&\vec{\lambda}_k^j = \vec{\lambda}_k^{j-1}, k = 1,\ldots, j-1,\\ &\vec{\lambda}_j^j = (1-s_j)\vec{e}_j\end{aligned} \tag{2.37}$$

$$\vec{\gamma}^j = \vec{\gamma}^{j-1} + \dot{q}_j(\vec{\omega}_j \times \vec{e}_j)(1-s_j), \tag{2.38}$$

$$\begin{aligned}&\vec{\xi}_k^j = \vec{\xi}_k^{j-1} - \vec{\lambda}_k^{j-1} \times \vec{r}_{j-1,j} + \vec{\lambda}_k^j \times \vec{r}'_{j,j}, k = 1,\ldots, j-1\\ &\vec{\xi}_j^j = \vec{e}_j s_j + \vec{\lambda}_j^j \times \vec{r}'_{j,j},\end{aligned} \tag{2.39}$$

$$\begin{aligned}&\vec{\delta}^j = \vec{\delta}^{j-1} - \vec{\gamma}^{j-1} \times \vec{r}_{j-1,j} + \vec{\gamma}^j \times \vec{r}'_{j,j} - \vec{\omega}_{j-1} \times\\ &\times(\vec{\omega}_{j-1} \times \vec{r}_{j-1,j}) + \vec{\omega}_j \times (\vec{\omega}_j \times \vec{r}'_{j,j}) + 2\vec{\omega}_j \times \vec{e}_j s_j \dot{q}_j\end{aligned} \tag{2.40}$$

In an iterative calculation, the j-th link is added to the chain in the j-th iteration and matrices Λ^j, Γ^j, Ξ^j, and Δ^j are calculated starting from Λ^{j-1}, Γ^{j-1}, Ξ^{j-1}, and Δ^{j-1}.

Among the robot links, special attention should be paid to the terminal one (index n), called *end effector*. Motion of the end effector is of key importance in the robot's task execution. This means that in practical robot operations, the spatial motion of the terminal link is prescribed and joint trajectories are the consequence. Therefore, it is necessary to define the position of the end effector by means of proper coordinates, called *external coordinates*. The end effector is a rigid body and its spatial position is described by means of six parameters. Three parameters define the position of the one point of the body, and the other three define the spatial orientation. The point that is positioned is selected depending on the nature of the task. Each task has one point of the end effector that is of main importance, called *tip*, and denote it by letter T (nozzle of the spray gun in a painting task, tip of the gripper, etc.). Figure 2.14a shows the Cartesian coordinates of the robot tip (x, y, and z) and the orientation angles: yaw, pitch, and roll (θ, φ, and ψ). For an exact definition we introduce a frame with the origin at the tip point and the axes parallel to the axes of the previously defined link-fixed frame of the end effector ($O_n x_n y_n z_n$ defined in Sec. 1.2). The new frame is $T x_n y_n z_n$ (orientation frame). Under transformation of the external fixed frame to the new one, we understand the translation of the origin (for the values of x, y, and z) and rotations for the angles θ, φ, and ψ (Fig. 2.14b). Therefore, the position of the end effector is defined by means of the 6-dimensional vector of external coordinates:

$$X = [x\ y\ z\ \theta\ \varphi\ \psi]^T \tag{2.41}$$

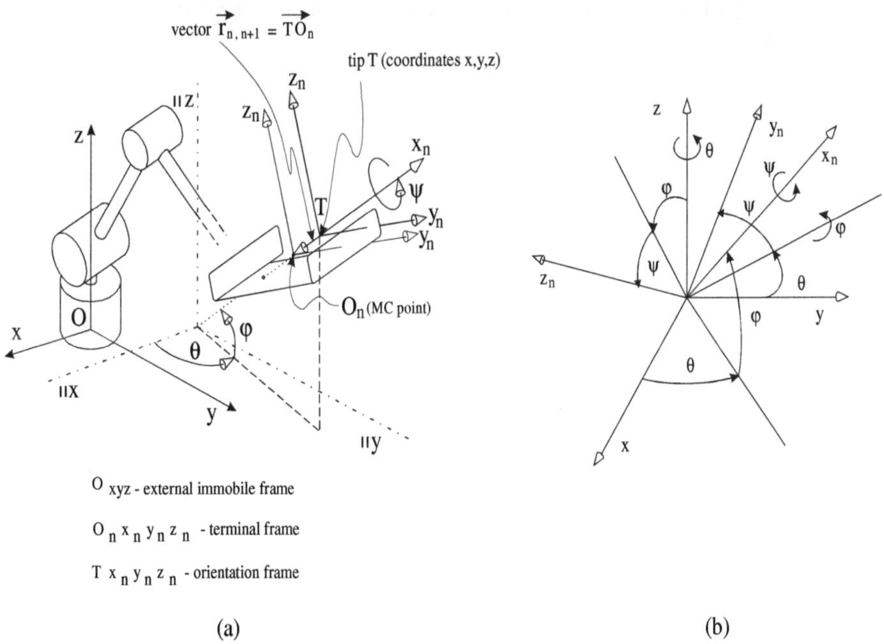

Figure 2.14. External coordinates describe the spatial position of the end effector: x, y, and z define the position of the tip, while the angles θ, φ, and ψ define the orientation.

It should be said that the axes x_n, y_n, and z_n may be oriented arbitrarily, but most often they are oriented along the principal axes of body inertia. If such directions are not proper for the definition of body orientation, a completely new frame may be introduced and used for orientation. This frame would be related to $O_n x_n y_n z_n$ by a constant transformation matrix.

We now discuss the relation between the internal position, n-dimensional vector q, and the external position, 6-dimensional vector X. We are especially interested in the possibility of establishing a unique correspondence between these two vectors. It is a rather complex problem and we shall give a brief and slightly simplified explanation. First, we note that different robot tasks require the different complexity of the external motion. Let n^* be the number of external coordinates needed for a particular task and let them form the reduced vector X^*.

- If $n < n^*$, there is no chance for a unique correspondence between q and X^* and the task cannot be executed. The vector X^* has at most n independent components.
- For n = n^*, there may be a unique correspondence. In most practical cases, the correspondence exists except for some positions which we call singular. In the text that follows we assume that the robot does not pass through singular positions. Thus, the equality of n and n^* ensures that the task can be executed.
- If $n > n^*$, we talk about redundancy. The correspondence between q and X^* does not exist. By its maneuvering ability, the robot overcomes the requirements imposed by the task.

We shall usually consider the problems where $n = n^* = 6$ and singularities do not exist. For this case, we search for the relation between q and X.

Let the tip T be defined relative to the end effector MC by using the vector $\vec{r}_{n,n+1}$ (Fig. 2.14a). This vector is prescribed among the other vectors of geometry by means of constant projections onto the link-fixed frame $\vec{\tilde{r}}_{n,n+1}$. Starting from (2.21) one obtains the expression for the external position of the tip:

$$(x, y, z) = \sum_{j=1}^{n} \vec{r}\,'_{j,j} - \vec{r}_{j,j+1}\,. \tag{2.42}$$

The orientation of the end effector is described by the transformation matrix between the external frame and the n-th link-fixed frame. On the one hand, the transformation matrix $A_n(q)$ can be solved, as explained in Sec.

1.2. However, the same matrix may be expressed in terms of orientation angles [5]:

$$A_n(\theta,\varphi,\psi) = A_\theta A_\varphi A_\psi \,, \tag{2.43}$$

where

$$A_\theta = \begin{bmatrix} \cos\theta & -\sin\theta & 0 \\ \sin\theta & \cos\theta & 0 \\ 0 & 0 & 1 \end{bmatrix}, \; A_\varphi = \begin{bmatrix} \cos\varphi & 0 & -\sin\varphi \\ 0 & 1 & 0 \\ \sin\varphi & 0 & \cos\varphi \end{bmatrix},$$
$$A_\psi = \begin{bmatrix} 1 & 0 & 0 \\ 0 & \cos\psi & -\sin\psi \\ 0 & \sin\psi & \cos\psi \end{bmatrix}. \tag{2.44}$$

Expressions (2.29)–(2.32), as well as the explained relations between the internal and the external coordinates (equations (2.42)–(2.44)), offer the possibility of formulating a Jacobian transform of the first or second order:

$$\dot{X} = J(q)\dot{q} \,, \tag{2.45}$$

$$\ddot{X} = J(q)\ddot{q} + A(q,\dot{q}) \,, \tag{2.46}$$

i.e. to calculate the *6x6* Jacobian matrix J and the 6×1 adjoint vector A.

Since we established the unique correspondence between q and X, it is possible to calculate one set of coordinates if the other set is given. Calculation of X for given q is called the direct problem of kinematics (or simply, direct kinematics), while the calculation of q for a given X is called the inverse problem of kinematics (inverse kinematics). Both problems can be successfully solved by using the Jacobian form (2.46).

2. DYNAMICS

Robot's motion is generated by the action of joint actuators. At this point we do not discuss the types of actuators but only state that a driving torque acts at each revolute joint, and a driving force at each linear joint. We use the term *drive* and the notation τ_j for the actuation of a joint j regardless of the joint type. The dynamics defines the relation between the joint drives and the chain motion, i.e. between the input and the output.

The first part of this section discusses the dynamics of an open-chain robot arm in the sense of drive–motion relation. The dynamic model is then expanded to encompass the dynamics of the actuators. Finally, we examine the effects of elasticity in the transmission system.

2.1 Dynamics of the Open-Chain Arm

Under the notion of dynamic model is understood a system of differential equations that describes robot dynamics. Among different approaches to dynamics, Newton–Euler's equations, Lagrange's equations, and Gauss' principle, we decided to present the first one, finding it most appropriate.

(a) Equations of dynamics

We consider the robot's dynamics starting with an individual link. Let it be the j-th one. For this purpose we fictitiously break the chain at the joints S_j and S_{j+1}. The disconnection at the joint S_j is shown in Fig. 2.15a.

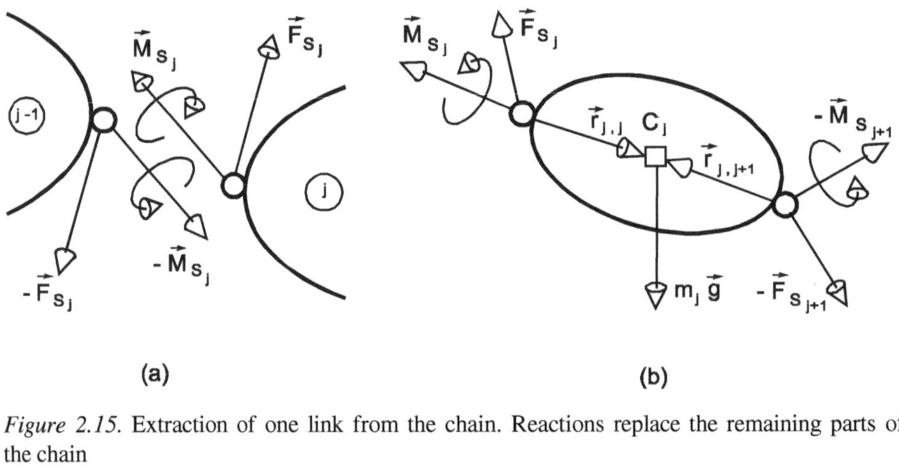

Figure 2.15. Extraction of one link from the chain. Reactions replace the remaining parts of the chain

The mutual influence of the two links is expressed in terms of a force and a couple. $\vec{F}_{Sj}$ denotes the force and $\vec{M}_{Sj}$ denotes the moment of the couple. They act in a forward direction (from the link j-1 to link j), while $-\vec{F}_{Sj}$ and $-\vec{M}_{Sj}$ act in a backward direction (from j to j-1).

Figure 2.15b shows the link j together with all forces and couples acting upon it. We use the law of linear momentum (Newton's law) to describe the motion of the MC:

$$m_j \vec{w}_j = m_j \vec{g} + \vec{F}_{Sj} - \vec{F}_{Sj+1} \, , \tag{2.47}$$

where m_j is the link mass, $\vec{w}_j$ is the MC acceleration, and $\vec{g}$ is the acceleration due to gravity: $(0, 0, -9.81\,[m/s^2])$. The law expresses the equilibrium of the real and inertial forces.

Now, we discuss the rotation of the link about its MC. It can be described by the law of angular momentum (Euler's equations):

$$\bar{J}_j\bar{\vec{\varepsilon}}_j + \bar{\vec{\omega}}_j \times (\bar{J}_j\bar{\vec{\omega}}_j) = \bar{\vec{M}}_{Sj} - \underline{\vec{M}}_{Sj+1} - \bar{\vec{r}}'_{j,j} \times \bar{\vec{F}}_{Sj} + \bar{\vec{r}}_{j,j+1} \times \underline{\vec{F}}_{Sj+1} \qquad (2.48)$$

$\bar{\vec{\varepsilon}}_j$ and $\bar{\vec{\omega}}_j$ are the angular acceleration and angular velocity. The bars above the letters or under the letters indicate that the vectors are expressed in the frame fixed to the link j. $\bar{J}_j$ is the tensor of inertia calculated for the axes of the link-fixed frame. In a general case, the tensor has the form

$$\bar{J}_j = \begin{bmatrix} J_{x_jx_j} & J_{x_jy_j} & J_{x_jz_j} \\ J_{y_jx_j} & J_{y_jy_j} & J_{y_jz_j} \\ J_{z_jx_j} & J_{z_jy_j} & J_{z_jz_j} \end{bmatrix} \qquad (2.49)$$

However, if the frame axes are placed so as to coincide with the principal inertial axes, then the tensor takes the diagonal form

$$\bar{J}_j = \begin{bmatrix} J_{x_j} & 0 & 0 \\ 0 & J_{y_j} & 0 \\ 0 & 0 & J_{z_j} \end{bmatrix}, \qquad (2.50)$$

where J_{x_j} is the moments of inertia with respect to the axis x_j and the analogous holds for J_{y_j} and J_{z_j}.

It should be mentioned that equation (2.47) is written in the external frame, while (2.48) is written in the link-fixed frame. Equation (2.47) could be simply rewritten and expressed in the link-fixed frame, while (2.48) cannot change the frame so easily. It would have to be multiplied by the transformation matrix A_j.

In the derived equations of dynamics, (2.47) and (2.48), there appear the force $\vec{F}_{Sj}$ and the couple $\vec{M}_{Sj}$, representing the total force and the total couple transmitted through the joint S_j. Let us examine these variables in more detail. The revolute joint is discussed first (Fig. 2.16a). It allows one relative motion, rotation about the axis $\vec{e}_j$. Hence, the reaction force $\vec{F}_{Rj}$

(from the link j-1 to link j) may have an arbitrary direction, while the reaction couple $\vec{M}_{Rj}$ is perpendicular to the axis $\vec{e}_j$ $(\vec{M}_{Rj} \perp \vec{e}_j)$. Besides the reactions, there exists a driving torque acting about $\vec{e}_j$. The drive is $\tau_j \vec{e}_j$, where τ_j is its value. If S_j is a linear joint (Fig. 2.16b), the reaction force $\vec{F}_{Rj}$ is perpendicular to the joint axis $(\vec{F}_{Rj} \perp \vec{e}_j)$, while the reaction couple $\vec{M}_{Rj}$ may have an arbitrary direction. Drive is now in the form of the force $\tau_j \vec{e}_j$, where τ_j is the force value. The total force and the total couple, for any type of joint, can be written as

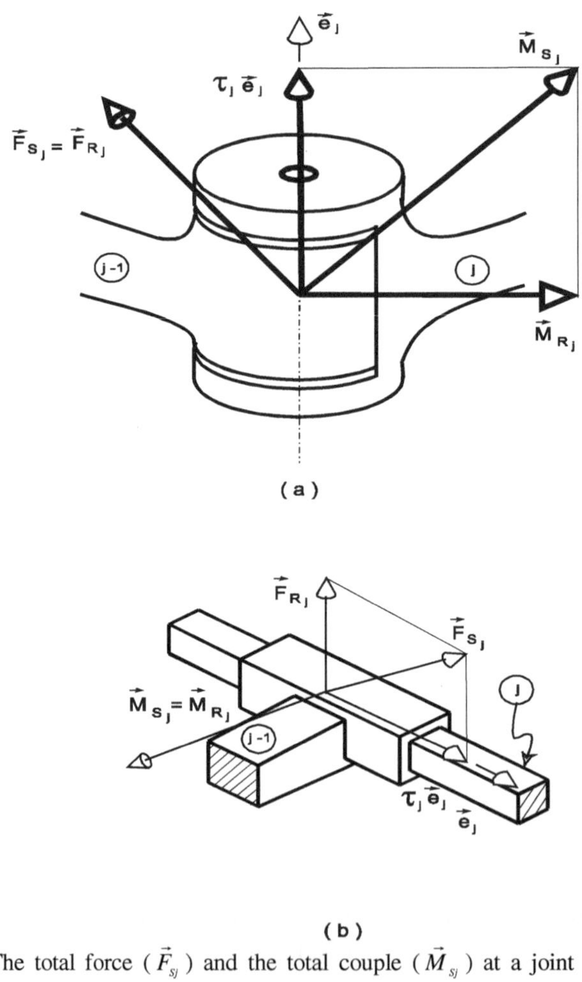

Figure 2.16. The total force ($\vec{F}_{Sj}$) and the total couple ($\vec{M}_{Sj}$) at a joint of the robot chain consist of reactions and the drive: (a) revolute joint, and (b) linear joint.

$$\vec{F}_{Sj} = \vec{F}_{Rj} + s_j \tau_j \vec{e}_j \,, \tag{2.51}$$

$$\vec{M}_{Sj} = \vec{M}_{Rj} + (1 - s_j) \tau_j \vec{e}_j \,, \tag{2.52}$$

where s_j is the joint type indicator.

Equations (2.47) and (2.48), together with relations (2.51) and (2.52), describe the dynamics of the link j. If the equations are written for all n links, a set of $2n$ vector equations is obtained:

$$m_j \vec{w}_j = m_j \vec{g} + \vec{F}_{Rj} + s_j \tau_j \vec{e}_j - \vec{F}_{Rj+1} - s_{j+1} \tau_{j+1} \vec{e}_{j+1} \,, \tag{2.53}$$

$$\begin{aligned} &\bar{J}_j \vec{\bar{\varepsilon}}_j + \vec{\bar{\omega}}_j \times (\bar{J}_j \vec{\bar{\omega}}_j) = \vec{\bar{M}}_{Rj} + (1 - s_j) \tau_j \vec{\bar{e}}_j - \underline{\vec{M}}_{Rj+1} - \\ &- (1 - s_{j+1}) \tau_{j+1} \underline{\vec{e}}_{j+1} - \vec{\bar{r}}'_{j,j} \times (\vec{\bar{F}}_{Rj} + s_j \tau_j \vec{\bar{e}}_j) + \\ &+ \vec{\bar{r}}_{j,j+1} \times (\underline{\vec{F}}_{Rj+1} + s_{j+1} \tau_{j+1} \underline{\vec{e}}_{j+1}), \\ &j = 1, \ldots, n, \end{aligned} \tag{2.54}$$

to describe the dynamics of the entire chain. We now search for a compact form of the dynamic model. Kinematic relations (2.25)–(2.28), along with (2.37)–(2.40), allow us to express the MC accelerations $(\vec{w}_j)$ and angular velocities and accelerations $(\vec{\bar{\omega}}_j, \vec{\bar{\varepsilon}}_j)$ in terms of internal coordinates q and their derivatives $\dot{q}, \ddot{q}$. Accelerations $\ddot{q}_1, \ldots, \ddot{q}_n$ are considered as unknowns in the dynamic model. The next step in the derivation of the model is the elimination of reactions. First, we check the number of equations against the number of unknowns in the systems of equations (2.53), (2.54). There are $2n$ vector equations or $6n$ scalar equations. On the other hand, there are two vector unknowns to be eliminated for each joint: $\vec{F}_{Rj}, \vec{M}_{Rj}, j = 1, \ldots, n$. However, these two vectors involve only five independent scalars. This is due to the fact that one of the reactions (force or torque) must be perpendicular to the joint axis. Thus, the entire number of scalar reactions is $5n$. If these reactions are eliminated from $6n$ equations there remains a set of n scalar equations that involve n scalar unknowns $\ddot{q}_1, \ldots, \ddot{q}_n$. It is a set of linear second-order differential equations:

$$\begin{aligned} &H_{11}(q)\ddot{q}_1 + \ldots + H_{1n}(q)\ddot{q}_n + h_1(q, \dot{q}) = \tau_1 \\ &\vdots \\ &H_{n1}(q)\ddot{q}_1 + \ldots + H_{nn}(q)\ddot{q}_n + h_n(q, \dot{q}) = \tau_n \end{aligned} \quad . \tag{2.55}$$

Notations $H_{ij}(q)$ and $h_j(q,\dot{q})$ express the dependence on the position and on the position and velocity, respectively. The model (2.55) can be written in a matrix form:

$$H(q)\ddot{q} + h(q,\dot{q}) = \tau \ , \tag{2.56}$$

where

$$\begin{aligned} &H = [H_{ij}]_{\dim = n\times n}, \ h = column\ [h_j], \\ &q = [q_1 \dots q_n]^T, \ \tau = [\tau_1 \dots \tau_n]^T. \end{aligned} \tag{2.57}$$

Matrix H is often called matrix of inertia. Vector h expresses the influence of the gravitational, centrifugal, and Coriolis' loads.

(b) Minimization of model form

We now describe the procedure for minimizing the model form, that is, for elimination of the reactions and calculation of the model matrices H and h.

The procedure starts with considering one part of the kinematic chain, from a selected joint to the free end (Fig. 2.17). Let the selected joint be S_j. We consider the subchain that consists of the links $j, j+1, \dots, n$. The fictitious break is made at the joint S_j and the influence of the preceding link is expressed in terms of the force $\vec{F}_{Sj}$ and the couple $\vec{M}_{Sj}$. Now, we apply D'Alembert's principle and establish the equilibrium of the real and the inertial forces acting on the considered part of the chain. The system of forces includes gravity forces, inertial forces, joint force $\vec{F}_{Sj}$, and joint couple $\vec{M}_{Sj}$.

The inertial load of a link (let it be the link k) is distributed all over it but can be reduced to a resultant inertial force acting at the MC, and a resultant inertial couple. The force is

$$\vec{F}_{kI} = -m_k \vec{w}_k \tag{2.58}$$

and it acts at the point C_k; m_k is the link mass. The moment of the inertial couple can be expressed in Euler's form

$$\vec{\bar{M}}_{kI} = -(\bar{J}_k \vec{\bar{\varepsilon}}_k + \vec{\bar{\omega}}_k \times (\bar{J}_k \vec{\bar{\omega}}_k)), \tag{2.59}$$

where J_k is the tensor of inertia. The couple (2.59) is expressed in the link-fixed frame and the transformation matrix A_k can be used to transfer it to the external frame: $\vec{M}_{kI} = A_k \vec{\overline{M}}_{kI}$. The gravity load of the link is

$$\vec{G}_k = m_k \vec{g} \,, \tag{2.60}$$

where $\vec{g}$ denotes the gravity acceleration.

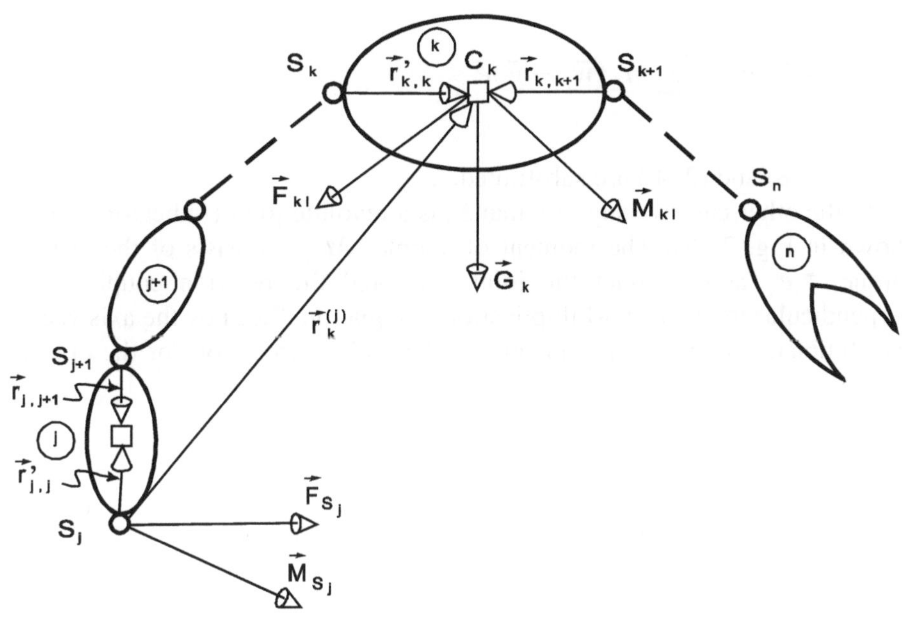

Figure 2.17. Part of the robot's chain.

D'Alembert's equilibrium concerns $\vec{F}_{kI}, \vec{M}_{kI}, \vec{G}_k, k = 1,\ldots,n, \vec{F}_{Sj}, \vec{M}_{Sj}$. Equilibrium of forces gives

$$\vec{F}_{Sj} = -\sum_{k=j}^{n} (\vec{G}_k + \vec{F}_{kI}) \,. \tag{2.61}$$

Equilibrium of moments of forces is found with respect to the point S_j yielding

$$\vec{M}_{Sj} = -\sum_{k=j}^{n} (\vec{M}_{kI} + \vec{r}_k^{(j)} \times (\vec{G}_k + \vec{F}_{kI})) \,, \tag{2.62}$$

where

$$\vec{r}_k^{\,(j)} = \overrightarrow{S_j C_k} = \sum_{p=j}^{k-1} (\vec{r}\,'_{p,p} - \vec{r}_{p,p+1}) + \vec{r}\,'_{k,k}\,. \tag{2.63}$$

Suppose now that the joint S_j is linear (indicator s_j=1), as shown in Fig. 2.16b. Force $\vec{F}_{Sj}$ consists of the driving component $\tau_j \vec{e}_j$ acting along the joint axis, and the reaction force $\vec{F}_{Rj}$ perpendicular to that axis. In order to eliminate the reaction, we multiply equation (2.61) by the joint axis vector $\vec{e}_j$, obtaining the scalar expression for the driving force

$$\tau_j = \vec{F}_{Sj} \vec{e}_j = \sum_{k=j}^{n} m_k (\vec{w}_k - \vec{g}) \vec{e}_j\,, \tag{2.64}$$

where (2.58) and (2.60) are substituted.

In the other case we suppose that S_j is a revolute joint (indicator s_j=0) as shown in Fig. 2.16a. The moment of couple, $\vec{M}_{Sj}$, consists of the driving torque $\tau_j \vec{e}_j$ acting about the joint axis, and the reaction couple $\vec{M}_{Rj}$ perpendicular to the axis. Multiplication of equation (2.62) by the axis vector $\vec{e}_j$ eliminates the reaction, thus giving the scalar expression for the driving torque

$$\begin{aligned} \tau_j &= \vec{M}_{Sj} \vec{e}_j = \\ &= \sum_{k=j}^{n} (A_k (\bar{J}_k \vec{\bar{\varepsilon}}_k + \vec{\bar{\omega}}_k \times (\bar{J}_k \vec{\bar{\omega}}_k)) + \vec{r}_k^{\,(j)} \times m_k (\vec{w}_k - \vec{g})) \vec{e}_j, \end{aligned} \tag{2.65}$$

where the expression (2.59) is multiplied by A_k, and expressions (2.58) and (2.60) are substituted.

To summarize, in the case of a linear joint the drive is expressed in the form (2.64) and in the case of a revolute joint in the form (2.65). This can be applied to the joints $j = 1,...,n$, and so, the dynamics of the robot chain is described by means of n scalar equations expressing the drives. In relations (2.64) and (2.65) there appear the MC accelerations, $\vec{w}_k$, and the angular accelerations, $\vec{\bar{\varepsilon}}_k$. These accelerations can be expressed in terms of joint accelerations by using the linear forms (2.32) and (2.31). In this way (2.64) is transformed to:

$$\tau_j = \sum_{k=j}^{n} e_j^T m_k (\Xi^k \ddot{q} + \Delta^k - g)\,, \tag{2.66}$$

where the vector notation is replaced by the 3×1 matrices. Further transformation yields

$$\tau_j = H'_j \ddot{q} + h'_j , \tag{2.67}$$

where

$$H'_j = \sum_{k=j}^{n} e_j^T m_k \Xi^k , \quad h'_j = \sum_{k=j}^{n} e_j^T m_k (\Delta^k - g) \tag{2.68}$$

are of dimensions 1×n and 1×1, respectively.

If (2.65) is transformed so as to depend on the joint accelerations, one obtains

$$\begin{aligned} \tau_j = & \sum_{k=j}^{n} e_j^T A_k (\overline{J}_k A_k^{-1} (\Lambda^k \ddot{q} + \Gamma^k) + \overline{\omega}_k^{\otimes} \overline{J}_k \overline{\omega}_k) + \\ & + e_j^T r_k^{(j)\otimes} m_k (\Xi^k \ddot{q} + \Delta^k - g) , \end{aligned} \tag{2.69}$$

where for any vector, e.g., $\vec{a}$, the notation $a^{\otimes}$ stands for the matrix

$$a^{\otimes} = \begin{bmatrix} 0 & -a_z & a_y \\ a_z & 0 & -a_x \\ -a_y & a_x & 0 \end{bmatrix} \tag{2.70}$$

that serves to present the vector product in a matrix form. Further transformation of (2.69) yields

$$\tau_j = H''_j \ddot{q} + h''_j , \tag{2.71}$$

where

$$\begin{aligned} H''_j &= \sum_{k=j}^{n} e_j^T (A_k \overline{J}_k A_k^{-1} \Lambda^k + r_k^{(j)\otimes} m_k \Xi^k), \\ h''_j &= \sum_{k=j}^{n} e_j^T (A_k \overline{J}_k A_k^{-1} \Gamma^k + A_k \, \overline{\omega}_k^{\otimes} \overline{J} \overline{\omega}_k + r_k^{(j)\otimes} m_k (\Delta^k - g)) \end{aligned} \tag{2.72}$$

are of dimensions 1 x n and 1 x 1, respectively.

In the way described we have found the expressions for the joint drives in terms of joint accelerations. For a linear joint, the form (2.67) holds along with (2.68), while for a revolute joint applies the form (2.71), along with (2.72). For the complete chain (all joints) the following matrix relation can be written

$$\tau = H\ddot{q} + h, \tag{2.73}$$

where

$$\tau = \begin{bmatrix} \tau_1 \\ \vdots \\ \tau_j \\ \vdots \\ \tau_n \end{bmatrix}, \quad H = \begin{bmatrix} H_1 \\ \vdots \\ H_j \\ \vdots \\ H_n \end{bmatrix}, \quad h = \begin{bmatrix} h_1 \\ \vdots \\ h_j \\ \vdots \\ h_n \end{bmatrix},$$

$$H_j \text{ and } h_j = \begin{Bmatrix} H'_j \text{ and } h'_j, & for \;\; s_j = 1 \\ H''_j \text{ and } h''_j, & for \;\; s_j = 0 \end{Bmatrix} \tag{2.74}$$

are of dimensions $\tau(n\times1)$, $H(n\times n)$, and $h(n\times1)$, respectively. Thus, we have come to the previously stated form of the dynamic model, that is, equation (2.56). Expressions (2.68) and (2.72) offer the possibility of recursive calculation of the matrices H and h.

Let us consider the solution of the dynamic model. The notions of the *direct* and the *inverse dynamics* are introduced in theoretical mechanics when considering the behavior of a mechanical system under the action of forces. In the direct dynamics problem, the motion of the system is known and the forces that cause this motion are to be calculated. The opposite problem, calculation of motion for the given forces, is called the inverse dynamics. One should note that some authors interchange these terms. Although one might find some justification for this, we adhere to the definition given above since "our" inverse dynamics needs the inversion of the inertial matrix, which gives the basis for an association. For a robotic chain, direct dynamics means calculation of the drives τ for the prescribed robot motion q(t), while in the inverse dynamics the motion q is calculated for the given drives τ. Model (2.56) offers the possibility of solving both problems.

Let us point out once again that τ_j represents the driving torque or the driving force, depending on the type of joint. In any case, it is an output variable of an actuator. This is an important conclusion since it implies that the model (2.56) describes the dynamics of the robot chain and does not take care of the dynamics of the actuation system and transmission. The model will be generalized in the next section.

2.2 Dynamics of Actuators and Transmission System

In this section we describe in brief the way of modeling actuator dynamics and the coupling of this model with the previously defined model of robot chain (mechanical structure). Effects of the actuator dynamics are demonstrated by considering a DC driven robot.

(a) Dynamics of a DC motor

Permanent magnet DC motors are very common actuators with robotic systems. Their main advantage is simple control by varying the input voltage. The main disadvantage, however, follows from the graphite brushes (except for brushless types). Some other problems, being common characteristics of all electric drives, should also be mentioned. It is a fact that they rotate quickly and produce relatively small torques. Thus, a gearbox is usually needed to reduce speed and increase torque. Further, motors are very often dislocated from the joints to the robot base in order to unload the arm statically. Hence, different transmissions (chains, belts, gears, shafts, etc.) are also needed. All this complicates the robot construction and may influence the accuracy of motion. However, good controllability makes DC motors still very popular.

The dynamics of a DC motor that drives the robot joint S_j is described by the following relations expressing the mechanical and electrical equilibrium:

$$I_j\ddot{\theta}_j = C_{Mj}i_j - B_j\dot{\theta}_j - M_j, \tag{2.75}$$

$$u_j = R_j i_j + L_j\, di_j/dt + C_{Ej}\dot{\theta}_j \tag{2.76}$$

where θ_j is the angle of motor shaft rotation, i_j is the armature current, M_j is the output torque, and u_j is the input voltage. Motor parameters are: I_j –rotor moment of inertia, C_{Mj} and C_{Ej} –constants of torque and back e.m.f., B_j – viscous friction coefficient, R_j –armature resistance, and L_j –inductivity. The dynamic equations can be united to obtain a more compact canonical form:

$$\dot{x}_j = C_j x_j + f_j M_j + d_j u_j, \tag{2.77}$$

where the state vector x_j has dimension three. The state vector and the system matrices are

$$x_j = \begin{bmatrix} \theta_j \\ \dot{\theta}_j \\ i_j \end{bmatrix}, \quad C_j = \begin{bmatrix} 0 & 1 & 0 \\ 0 & -B/I & C_M/I \\ 0 & -C_E/L & -R/L \end{bmatrix},$$
$$f_j = \begin{bmatrix} 0 \\ -1/I \\ 0 \end{bmatrix}, \quad d_j = \begin{bmatrix} 0 \\ 0 \\ 1/L \end{bmatrix}, \tag{2.78}$$

where, for simplicity, the index j is omitted from the elements of system matrices.

If the inductivity L is small enough (it is a rather common case), then the term Ldi/dt can be neglected. Equation (2.76) now becomes

$$u_j = R_j i_j + C_{Ej}\dot{\theta}_j, \tag{2.79}$$

and the number of state variables reduces to two. The state vector and the system matrices in equation (2.77) are

$$x_j = \begin{bmatrix} \theta_j \\ \dot{\theta}_j \end{bmatrix}, \quad C_j = \begin{bmatrix} 0 & 0 \\ 0 & -C_M C_E/RI - B/I \end{bmatrix},$$
$$f_j = \begin{bmatrix} 0 \\ -1/I \end{bmatrix}, \quad d_j = \begin{bmatrix} 0 \\ C_M/RI \end{bmatrix}. \tag{2.80}$$

It was stated that DC motors are usually accompanied by some transmission system. The transmission defines the relation between the joint coordinate q_j and the motor coordinate θ_j. If the transmission is considered ideal (no backlash, no elastic deformation) then the transmission ratio is constant:

$$q_j = \theta_j / N_j . \tag{2.81}$$

If the dynamics of the transmission system (inertial properties) and the loss due to friction are neglected, then it holds that

$$\tau_j = M_j N_j . \tag{2.82}$$

If friction has to be discussed, then it is considered through the power loss. The efficiency coefficient is introduced: $0 \le \eta_j \le 1$. Now, equation (2.82) is modified. If the motion is in the direction of the drive, then $N_j \eta'_j$ is used instead of N_j. However, if the motion is opposite to the action of the drive, then N_j / η''_j applies. Note that η'_j and η''_j are generally different. The efficiency of a gearbox in the reverse direction is smaller $(\eta''_j < \eta'_j)$. If inertial effects of transmission should be taken into consideration, they can be easily added to the motor inertia I_j.

(b) The complete dynamic model

General discussion on the interaction of the chain dynamics and actuators dynamics would be based on the model (2.55) for the chain and model (2.77) along with (2.81) and (2.82) for actuators and transmission [4, 6, 7]. Here, we present a simplified approach and stress only the main dynamic effects. In the actuator model, the inductivity L is neglected and also the viscous friction B (real numerical values justify this approximation). Equations (2.75) and (2.76) now yield

$$I_j \ddot{\theta}_j = \frac{C_{Mj}}{R_j} u_j - \frac{C_{Mj} C_{Ej}}{R_j} \dot{\theta}_j - M_j . \tag{2.83}$$

Relations (2.81) and (2.82) that describe the transmission are needed to connect variables of the motor and of the joint. The motor variables in equation (2.83), θ and M, are replaced by the joint variables, q and τ. The torque τ is then substituted from such modified relation (2.83) into the system (2.55), thus yielding the model

$$\begin{aligned} &H_{j1}\ddot{q}_1 + \ldots + (H_{jj} + I_j N_j^2)\ddot{q}_j + \ldots + H_{jn}\ddot{q}_n + h_j + \\ &+ (C_{Mj} C_{Ej} / R_j) N_j^2 \dot{q}_j = (N_j C_{Mj} / R_j) u_j \ , \quad j = 1, \ldots, n, \end{aligned} \tag{2.84}$$

or in the matrix form

$$H^*(q)\ddot{q} + h^*(q, \dot{q}) = Du \, , \tag{2.85}$$

where

$$H_{kj}^* = \begin{Bmatrix} H_{kj}, & \text{for } k \neq j \\ H_{jj} + I_j N_j^2, & \text{for } k = j \end{Bmatrix},$$

$$h_j^* = h_j + (C_{Mj} C_{Ej} / R_j) N_j^2 \dot{q}_j, \tag{2.86}$$

$$D = diag[N_j C_{Mj} / R_j].$$

By comparing the models (2.55) and (2.84) one may conclude that the introduction of motor dynamics reflects on the increased proper inertia of the joints (the diagonal coefficients H_{jj} are augmented), while the dynamic coupling between the joints remains the same (the nondiagonal coefficients H_{kj} do not change).

The discussion of actuators could be rather complex since robots might be driven by different types of actuators: electrical, hydraulic, or pneumatic [7]. In order to achieve a uniform approach to actuator dynamics, we describe it in a canonical form (2.77). It holds for any type of actuator but the system matrices and the order of the system differ. For a hydraulic cylinder with a servovalve, the discussion on dynamics usually starts with a fifth-order model. Under some assumptions it can be reduced to third-order. As explained in this section, DC motor can be modeled by using a third-order form, or the order may be reduced to two. We have demonstrated how to handle the second-order model. Any higher-order model needs a more complex treatment [6, 7].

2.3 Dynamics of Arm with Flexible Transmissions

Here we consider an open-chain mechanism and concentrate on the problem of transmitting torques from the motors to joints. With the robots driven by electrical motors, it is very common that the motor is displaced from the joint. They are dislocated to the robot base in order to achieve better statics (that is, to unload the arm). In such case, different transmissions can be applied between the motor shaft and joint shaft. Some kind of a gearbox is usually present and accompanied by chains, belts, or some other elements. Such transmission is a source of deviation since each element may introduce its elastic deformation. Here, we assume that each joint has only one deformable element in the transmission. In order to make the discussion clearer, let the elastic element be the Harmonic-Drive reducer (HD). This will not compromise the generality since any other transmission element may be modeled in the same manner. A general discussion may be found in [8]. Fig. 2.18a shows the joint S_j with its motor and transmission.

Motor dynamics can be described by the mechanical equation (2.75) and the electrical equation (2.76). If we neglect friction and inductivity then these two equations can be combined to give the form (2.83). This model is expressed in terms of the motor shaft angle θ_j. M_j which appears in the model, represents the motor output torque. The model holds for each motor, j=1, ... , n.

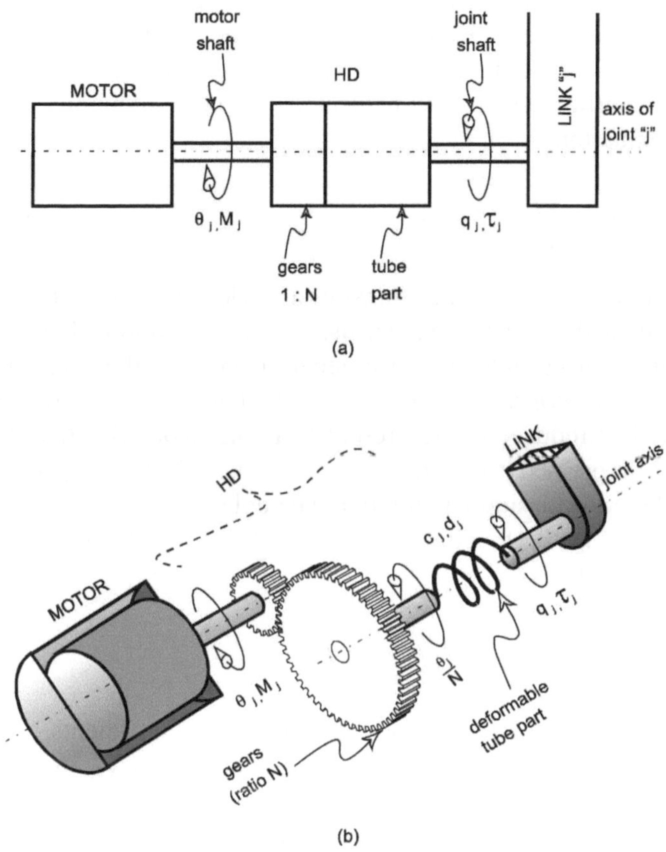

Figure 2.18. Deformable transmission

Dynamics of the robot links is described by the model (2.56) and expressed in terms of the joint coordinates q_j, j=1, ... , n. Vector $\tau = [\tau_1 \ldots \tau_n]^T$ which appears in the model contains the input torques for the robot links (joint shaft torques).

In the preceding paragraph, when nondeformable transmission was assumed, the relation between the motor variables (θ_j, M_j) and joint variables (q_j, τ_j) was defined as linear. It was expressed by means of equations (2.81) and (2.82). Motion of the motor and that of the corresponding joint were dependent on each other. If flexibility of transmission is introduced, the situation becomes rather different. The motion of the motor (coordinate θ_j) and the motion of the joint (coordinate q_j) become kinematically independent, and only the high stiffness keeps q_j close to θ_j/N_j. This means that the overall number of DOFs is doubled. We now concentrate on the concrete example of transmission shown in Fig. 2.18. In order to make the discussion more general, besides the elasticity we shall take care of the transmission inertia. In the schematic presentation (Fig. 2.18b), the dashed line indicates the transmission stage. Its input torque is M_j and the output is τ_j. If the moments of inertia of the gears are I'_j and I''_j and if other inertial effects are neglected, then the dynamics can be described by the equation

$$(I'_j + I''_j/N_j^2)\ddot{\theta}_j = M_j - \tau_j/N_j . \tag{2.87}$$

It should be noted that any other transmission element (chain, belt, etc.) could be modeled in the same way, that is, as a combination of two rigid gears and a torsion spring element. In a generalized case, the transmission may consist of several consecutive stages. Each stage would be treated as explained for the HD reducer in the present example. Note also that inertia of gears may be expressed to a stronger or weaker degree, especially if compared with the transmission friction that we neglected.

Let us return to Fig. 2.18. The joint torque τ_j follows from the spring deformation, and hence

$$\tau_j = c_j(\theta_j/N_j - q_j) + d_j(\dot{\theta}_j/N_j - \dot{q}_j), \tag{2.88}$$

where c_j and d_j denote the torsion stiffness and damping coefficients.

The complete dynamic model now includes equations (2.56) for the links, (2.83), $j=1,...,n$ for the motors, and (2.87), (2.88), $j=1,...,n$ for the transmissions. The model can be integrated to give the motions θ_j and $q_j, j=1,...,n$.

3. CONTROL CONCEPTS – MAIN IDEAS

Although this book is concerned with robot dynamics, and has no intention of discussing the problems of robot control, we still feel it necessary to explain some basic ideas. The problems of control are so involved with the system dynamics, that the discussion of dynamics would not be complete without including some issues on control. In fact, one reason for studying dynamics is to improve the knowledge about the system in order to better control it.

The first question that we face is how to define the notion of *control problem*. First, we have to think about what the robot is expected to do. It is usually said that the robot has to perform a *prescribed task*. However, the definition of a task may be rather different. In a most simple case the task is considered as executing prescribed motion – *kinematically-oriented task*. In a more complex case the task may include regulation of some dynamic variables, such as forces, etc. (*dynamically-oriented task*). Finally, the generalization of the task notion may lead to the *problem-oriented* definition. For example, one may state the task as follows: "assemble the given device". In this situation it is necessary to apply artificial intelligence to "translate" the task to kinematic and dynamic forms.

In this chapter we discuss the dynamics of open-chain robots and accordingly the control problem will be considered in its simplest form: *find and apply the control input that will ensure execution of the prescribed motion*. Since the robot motion can be described by two different sets of coordinates, the robot task may prescribe the time histories of the internal coordinates $q^*(t)$ or the external coordinates $X^*(t)$. Symbol * indicates that the required motion is in question. We also use the terms reference motion, desired motion, or nominal motion. The same symbol will be used with other variables for their reference values (e.g., reference force). Thus, in our simplified task the input to a hypothetical control unit is the reference motion, and the output is the control signal u that is applied to the robot actuators. In order to handle both the internal and external coordinates, a two-level control unit is proposed, as shown in Figure 2.19. The goal of the *kinematic level* is to solve the inverse kinematics, i.e. to calculate the reference internal motion $q^*(t)$ for the given reference external motion $X^*(t)$. The problem of kinematics was discussed in section 1.3. Here, we concentrate on the *executive level* of control.

On the executive level we relate the reference motion $q^*(t)$ and the control u. In order to achieve the desired quality of tracking, it is necessary to ensure information about the real motion. Hence, sensors are used to measure the real motion q. Now, the reference motion and the real motion represent the basis for the calculation of control (closed-loop control scheme). This is what the control unit has to do. The described problem belongs to the class of nonlinear control. The complex theory of nonlinear control is not the topic of this book; we will explain here only the main ideas.

The dynamic model that was discussed in previous sections relates the motion and the control (see equation (2.85)). So, it is possible to start from the reference motion $q^*(t)$ and calculate the corresponding control. Such control, obtained from the dynamic model, is called the *reference control* and is denoted by u^* (the term nominal control is often used). In an *open-loop* control system, the reference control is to be applied to the robot actuators. However, it is necessary to add the correction term in order to compensate for the unmodeled dynamic effects and different external perturbations. Hence, the control that is applied to actuators has two terms:

$$u = u^* + \Delta u \,. \tag{2.89}$$

The term u^* (reference control) represents the *feedforward* signal, while the correction term Δu is *feedback*. The feedforward signal may be calculated in advance (off-line), that is before the task execution. It is then stored in memory and used when required. This means that the task is completely defined in advance, and this is usually the case with standard industrial tasks. However, a more advanced approach should handle the situation when the task is prescribed on-line (e.g. joystick guiding or sensor guiding). In such case, the calculation of u^* is performed on-line. This assumes that the dynamic model (direct dynamics) can be solved in real time.

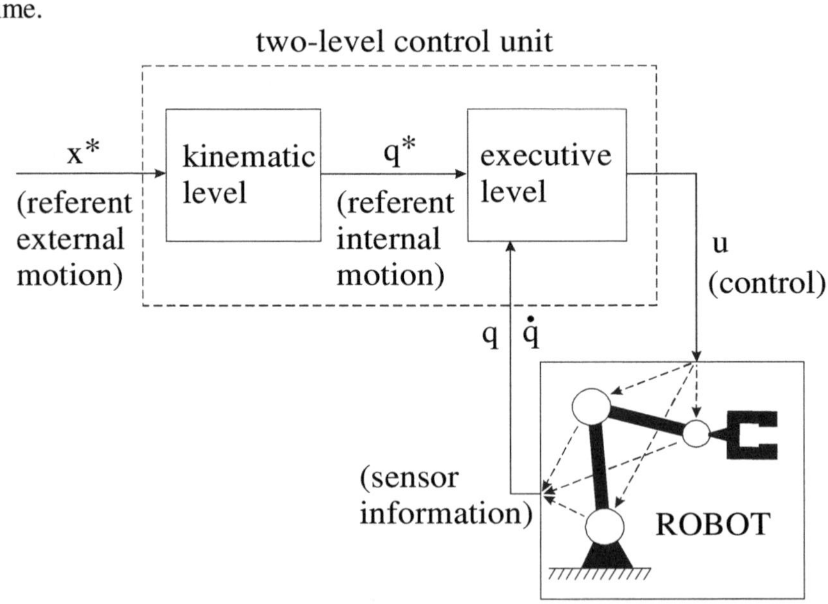

Figure 2.19. Two-level robot control unit

The feedback Δu is solved on the basis of the motion error $\Delta q = q - q^*$. It is always calculated on-line.

In most practical realizations, the control law is simplified by omitting the feedforward part. The control (2.89) is reduced to $u = \Delta u$. Furthermore, the feedback is simplified by decoupling the system into subsystems. Now each actuator (that is each robot joint, e.g., the j-th one) represents a separate system and is controlled independently by applying the local linear regulator:

$$\Delta u_j = -K_j^p \Delta q - K_j^d \Delta \dot{q} - K_j^i \int \Delta q \, dt \tag{2.90}$$

where K_j^p, K_j^d and K_j^i are the proportional, differential, and integral feedback gains respectively. In [6], it was shown that local control could not guarantee stability of the entire system. So-called *global control* term was necessary in some examples. However, this is not done in practice.

The advanced control units that should support fast and precise manipulation use feedforward and apply the control law (2.89). In this way, they take into account the coupling of the subsystems (feedforward is calculated from the model of the complete system). Feedback is still kept in the form (2.90).

REFERENCES

1. Yu. Stepanenko, "Method of Analysis of Spatial Lever Mechanisms", (in Russian), Mekhanika Mashin, Vol. 23, Moscow, 1970.
2. M. Vukobratovic, Yu. Stepanenko, "Mathematical Models of General Anthropo-morphic Systems", Math. Biosciences, Vol. 17, pp. 191-242, 1973.
3. J. Denavit, R. S. Hartenberg, “A Kinematic Notation for Lower-Pair Mechanisms Based on Matrices”, J. Applied Mechanics, pp. 215-221, June 1955.
4. M. Vukobratovic, V. Potkonjak, Scientific Fundamentals of Robotics, Vol. 1: Dynamics of Manipulation Robots, Springer-Verlag, 1982.
5. M. Vukobratovic, V. Potkonjak, Scientific Fundamentals of Robotics, Vol. 6: Applied Dynamics and CAD of Manipulation Robots, Springer-Verlag, 1985.
6. M. Vukobratovic, D. Stokic, Scientific Fundamentals of Robotics, Vol. 2: Control of Manipulation Robots, Springer-Verlag, 1982.
7. M. Vukobratovic, Applied Dynamics of Manipulation Robots: Modeling, Analysis and Examples, Springer-Verlag, 1989.
8. V. Potkonjak, "Contribution to the Dynamics and Control of Robots Having Elastic Transmissions", Robotica, Vol. 6, pp. 63-69, 1988.
9. N. Kircanski, A. Timchenko, M. Vukobratovic, “Position Control of Robot Manipulators with Elastic Joints Using Force Feedback”, Journal of Robotic Systems, Vol. 7, pp. 535-545, 1990.

Chapter 3

RIGID-BODY CONTACT OF A ROBOT WITH ITS ENVIRONMENT

In this chapter we discuss the contact tasks in which the robot environment is considered as a geometric constraint imposed to the motion of the end-effector. The first general approach to this problem was presented in [1].

1. MATHEMATICAL FORMULATION OF CONTACT TASK

1.1 Kinematics and Dynamics: Constraints, Degrees of Freedom, and Reactions

The discussion starts with the free motion of a rigid-body robot. The position of the chain, consisting of n links and n one-DOF joints (Fig. 2.5), is defined by means of the joint coordinates: $q = [q_1 \ldots q_n]^T$. In practical operations, joint coordinates are not always suitable for use. Many tasks can not easily be described in this way. For this reason, a new set of coordinates is defined – the external coordinates. By this term we usually understand position and orientation of the end-effector with respect to the immobile frame: Cartesian coordinates of the robot tip (x, y, z) and yaw, pitch, and roll angles (θ, φ, ψ), as shown in Fig. 2.14. The complete external position vector is $X = [x\ y\ z\ \theta\ \varphi\ \psi]^T$ and has a dimension of six. For simplicity, we assume that the robot has six joints and, thus, six joint coordinates $(n = 6)$. In this way the elaboration of redundancy is avoided but the intention of our discussion is not compromised. In the majority of manipulation tasks, the external position is very convenient to apply. However, with contact tasks and some other process operation, it is more appropriate if the position of the end-effector is defined relative to the object being processed. Hence, we introduce a frame fixed to the object. Since the object might be mobile, the new frame can also move. We assume that the object moves according to a given law that cannot be affected by the robot's action. This is necessary if one wants to describe the environment as a geometric constraint. With respect to the new frame, the position and the orientation of the end-effector are expressed by means of six coordinates: $s = [s_1 \ldots s_6]^T$. We call them *functional coordinates* [2,3]. Two examples are shown in Fig. 3.1. Case (a) represents the surface-type constraint appropriate for modeling the writing task. Case (b) represents the peg-in-hole assembly task. Note that the order of coordinates in the vector s may be chosen arbitrarily and we adopt the order suitable for the discussion that follows.

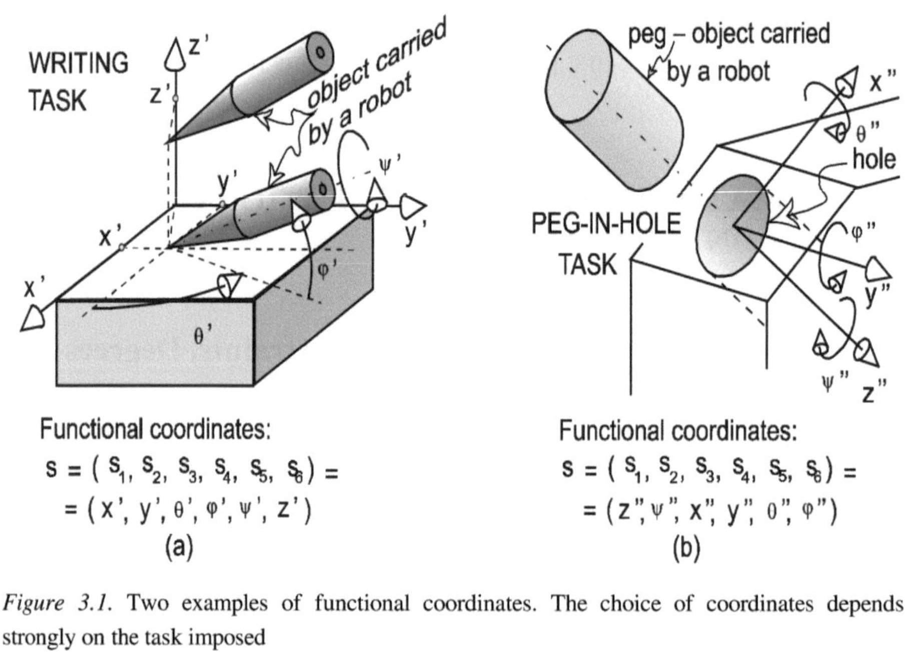

Figure 3.1. Two examples of functional coordinates. The choice of coordinates depends strongly on the task imposed

Each contact task consists of three phases: approaching, impact, and constrained motion. In the approaching phase, the end-effector moves towards the constraint. The motion is usually planned so as to achieve a zero-velocity contact and avoid impact. However, in a real situation, the omnipresent perturbations cause collision with the constraint. The impact occurs, producing sudden change in velocities. After the impact phase, the regular constrained motion starts. From the standpoint of modeling, the first phase represents free motion. Both, the internal coordinates (q) and the functional coordinates (s) are free. All the discussion from the previous chapters holds and the dynamics is described by the model (2.56). The kinematics should be described by means of s-coordinates. The relation between the coordinates q and s can be expressed via a nonlinear function that can be nonstationary or stationary:

$$s = s(q,t) \quad \text{or} \quad s = s(q) \tag{3.1}$$

With the nonstationary problems, explicit appearance of time t is due to the mobility of the s-frame. For reasons of simplicity we shall confine our discussion to stationary problems (immobile frame s). The second derivative of the latter relation from (3.1) produces

$$\ddot{s} = J(q)\ddot{q} + A(q,\dot{q}) \tag{3.2}$$

where $J = \partial s/\partial q$ is the Jacobian matrix, and $A = (\partial^2 s/\partial q^2)\dot{q}^2$ is the adjoint vector.

Although all the coordinates from the set s are free and independent (in the approaching phase), it is useful to partition them into two subsets, s^f and s^c, of dimensions *6-m* and *m*, respectively: $s = [s^{f^T} s^{c^T}]^T$. This partition follows from the nature of the constraint being approached, and will be explained later. With the vector s decomposed in two subvectors, the relation (3.2) becomes

$$\ddot{s}^f = J^f(q)\ddot{q} + A^f(q,\dot{q}) \tag{3.3}$$

$$\ddot{s}^c = J^c(q)\ddot{q} + A^c(q,\dot{q}) \tag{3.4}$$

where the Jacobian and the adjoint vectors are decomposed into blocks: J^f and A^f contain the upper 6-*m* rows, while J^c and A^c contain the rest (the lower *m* rows). Dimensions of the blocks are: $J^f((6-m)\times 6)$, $A^f(6-m)$, $J^c(m\times 6)$, and $A^c(m)$.

The approaching phase ends when the end-effector touches the constraint, that is, when the corresponding coordinate becomes zero. For the example from Fig. 3.1a, the contact is established when the coordinate s_6 (that is, z') reduces to zero. The impact occurs at the instant of contact. We find it more convenient to discuss impact dynamics later and elaborate now the constrained motion (the third phase). In this case, we assume that the contact has already been established and all the effects characteristic of the impact process ceased. The imposed geometric constraint restricts a certain motion of the end-effector. Let this restriction be expressed in terms of the *s*-coordinates by means of the *m*-dimensional condition

$$s^c = 0 \; . \tag{3.5}$$

For example (a) from Fig. 3.1, the decomposition of the vector *s* should be: $s^f = [s_1 \ldots s_5]^T$, $s^c = [s_6]$, $m = 1$; while in example (b) the decomposition: $s^f = [s_1 \; s_2]^T$, $s^c = [s_3 \; s_4 \; s_5 \; s_6]^T$, $m = 4$, is applied. Thus, s^c represents the set of constrained coordinates (the superscript "*c*" stands for "*constrained*"). In example (a) there is only one such coordinate and accordingly one DOF is lost while five remained. In example (b), four DOFs are lost and two remained. In a general case, the constrained robot has $6-m$ remaining DOFs and its motion is described by means of $6-m$ coordinates forming the vector s^f (the superscript "*f*" stands for "*free*").

The restriction of motion results in the reaction forces. Reactions will appear in the directions of the constrained coordinates s^c. Thus, there will be *m* independent components of the reaction. Let these components form the vector F. The reaction component may be in the form of force if translation is constrained, or in the form of torque if rotation is constrained. Fig. 3.2 shows the reactions for the examples defined in Fig. 3.1.

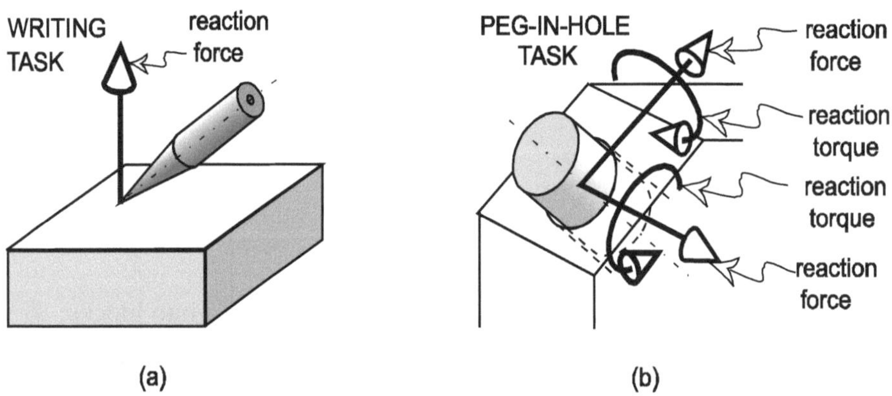

Figure 3.2. Two examples of reactions. Reactions correspond to the contact tasks shown in Fig. 3.1.

The dynamic model (2.56) used to describe free motion should now be reformulated in order to incorporate the reaction effects. Since F is the external force acting in the directions s^c, the model can be written in the form:

$$H(q)\ddot{q} + h(q,\dot{q}) = \tau + J^{c^T}(q)F \tag{3.6}$$

where $J^c = \partial s^c / \partial q$ was introduced in (3.4). In order to solve the dynamics of the constrained motion, the model (3.6) has to be considered along with the condition (3.5). It is more suitable if the constraint is expressed in differential form, and hence, (3.4) should be used, yielding

$$\ddot{s}^c = J^c(q)\ddot{q} + A^c(q,\dot{q}) = 0 \tag{3.7}$$

The matrix equations (3.6) and (3.7) complete the dynamic model. They involve $6+m$ scalar equations and can be solved for $6+m$ scalar unknowns: accelerations $\ddot{q}$ (6 components) and reactions F (m components). Calculation of the accelerations $\ddot{q}$ enables simulation of the system dynamics.

Most of the expressions derived above concern stationary problems. If a nonstationary constraint is imposed then the first relation from (3.1) has to be used. The second derivative of the coordinates s still has the general form (3.2) but the system matrices are different. The Jacobian keeps the original definition but features explicit dependence on time: $J(q,t) = \partial s / \partial t$. The adjoint vector is now rather different: $A(q,\dot{q},t) =$ $= (\partial^2 s/\partial q^2)\dot{q}^2 + 2(\partial^2 s/\partial q\,\partial t)\dot{q} + \partial^2 s/\partial t^2$. These differences are reflected in all other expressions in the dynamic model.

However, in the discussion that follows we shall mainly consider stationary problems. The differences that come from nonstationary effects will be referred to when necessary.

1.2 Impact

Now, we return to the second phase of the contact task, the problem of impact. At the beginning, we consider the simple surface-type constraint shown in Figs. 3.1a and 3.2a. The impact occurs when s^c (that is, the coordinate s_6) becomes zero and the impact action is directed along this coordinate. The impact can be characterized by a different degree of elasticity and, depending on that, skipping can appear. In order to avoid a complex discussion, we restrict the consideration to the nonelastic problem. In this case, the end-effector will not leave the surface after the first contact but will move along it.

A slightly more complex problem would be the constraint in the form of two surfaces. Although the intention is to move the end-effector along the line of intersection, in a real situation, the collision with one surface will happen first. Thus, two impacts will occur, one after the other. However, one might neglect this effect and consider two contacts as simultaneous. In this case a complex impact has two components acting along the two constrained coordinates (dimension of s^c is 2).

Further generalization leads to the *m*-component constraint, that is, *m* restricted coordinates of the vector s^c. The complex impact force F will have *m* components acting simultaneously. The example for $m = 4$ is shown in Figs. 3.1b and 3.2b. Note that in a real motion the complex impact represents a series of collisions with surfaces, but in the discussion that follows we shall neglect this effect.

In order to solve the impact dynamics, we integrate the model of constrained motion over the impact interval. Let t' denote the beginning of the impact, that is, the instant when s^c reduces to zero and the contact occurs. Let t'' be the instant when the impact process may be considered finished. The impact interval is then $\Delta t = [t', t'']$. For a geometric constraint and a nonelastic impact, it is justified to consider this interval infinitely short, that is, $\Delta t \to 0$ and $t'' \to t'$. We integrate the model (3.6) over this interval to obtain:

$$H(q')\Delta\dot{q} = J^{c^T}(q')F\Delta t \ , \ \ \Delta\dot{q} = \dot{q}'' - \dot{q}' \tag{3.8}$$

where $\dot{q}' = \dot{q}(t')$ and $\dot{q}'' = \dot{q}(t'')$. Then, equation (3.7) yields:

$$J^c(q')\dot{q}'' = 0 \Rightarrow J^c(q')\Delta\dot{q} = -J^c(q')\dot{q}' \tag{3.9}$$

Relations (3.8) and (3.9) describe dynamics of the impact. They represent the set of 6+*m* scalar equations and can be solved for $\Delta\dot{q}$ and $F\Delta t$. $\Delta\dot{q}$ is the change in joint velocity (6 components) and $F\Delta t$ is the impact impulse (*m* components). Note that $F\Delta t \neq 0$ although $\Delta t \to 0$. This means that $F \to \infty$. One should mention that the position does not change during the infinitely short impact, that is $q' = q''$. Hence, for the matrices H and J^c, it is not necessary to indicate whether they are calculated for q' or for q''.

Let us explain in brief how we use the dynamic models to solve numerically the three phases of the contact task. In the approaching phase, the model (2.56) is integrated checking constantly the value of s^c. When s^c becomes zero, we turn to the impact model (3.8), (3.9) and solve the change in velocity. With this new initial state we start solving the dynamics of the constrained motion by integrating the model (3.6), (3.7).

A few additional comments should be made about the impact. We have stated that the approaching phase ends when s^c becomes zero. At that instant the collision with the constraint occurs. In our discussion of impact we have assumed that the impact was infinitely short $(\Delta t \to 0)$, and after that "regular" constrained motion started. However, the fact is that some time after the instant of collision the motion features some specificity. There are often large variations in the reaction force, errors in the trajectory tracking, etc. These features are the result of the impact shock. So, it might be justified to say that impact consequences can be observed for some time after the collision (although the impact in a "classical" sense is considered infinitely short and the velocities change instantly). For this reason we often use the term *transient phase* to describe the time when the motion is still not completely "regular". This will be shown in the examples.

1.3 Some Control Issues

Control problems in contact tasks are much more complex then with free motion. With a free robot there existed n DOFs, that is, n independent coordinates to control. For simplicity reasons, let us assume that $n = 6$. If constraints are imposed, the number of independent coordinates (that is, the number of DOFs) reduces. In a general case, constraints will restrict m coordinates making them dependent on the remaining $6-m$. Thus, $6-m$ free coordinates remain to control. At first sight, it looks like the control problem is simplified. On the kinematic level, this conclusion might be justified. However, dynamics gives the insight into the true nature of the problem. Restriction of some coordinate is followed by the introduction of a reaction force (or couple). We often use the term contact force. If m DOFs are restricted, the m-dimensional reaction appears. If the control problem is formulated as pure kinematic – request that the robot tracks a given reference motion – it is likely that reaction forces will vary greatly, taking even extremely high values. In practice, constrained tasks refer to writing, some process operations (e.g., deburring), and assembly. In any of these tasks, strong contact forces may damage the contact surfaces. Hence, it appears necessary to control the reactions. Moreover, in many tasks it is not enough to prevent high forces but they have to be kept at prescribed values. Then we talk about tracking the reference forces. Now, we are again in a position to control $n = 6$ variables: $6-m$ coordinates and m contact forces. Control of forces makes the problem complicated. If contact is modeled by means of geometric constraints, then the contact forces cannot be expressed as algebraic functions of the state variables $q, \dot{q}$. Forces include the second derivative – acceleration $\ddot{q}$. For this problem, a new approach is needed.

Control inputs may be based on the feedforward-plus-feedback principle stated in Section 3 in Ch. 2. Expression (2.89), that is

$$u = u^* + \Delta u,$$

can still be used. The term u^* is the so-called reference (or nominal) control, calculated from the dynamic model. For constrained motion, the model (3.6), (3.7) is applied. Reference control represents the feedforward signal. Calculation of the feedback Δu is a more complex problem that will be discussed later.

The intention of this book is not to elaborate the control issues in detail. However, it is still necessary to discuss the principles of control. We do this in a way that explains the matter briefly, yet clearly and systematically. Hence, the book is supplemented with two appendices (Appendix 1. and Appendix 2.) that give a systematic overview of the control concepts for contact tasks.

2. WRITING TASK – ROBOT ARM SUBJECT TO SURFACE-TYPE CONSTRAINT

Let us consider a robot with the imposed constraint upon the motion of the tip of its end-effector. We restrict the tip by forcing it to move along a prescribed surface. This is called a surface-type constraint. Note that we are not dealing with the general case. The end-effector must not roll over the surface but one specific point, the tip, must always be in contact. This type of constraint reduces the number of DOFs by one ($m = 1$, according to the notation from Section 1.1). Thus, five DOFs of the end-effector remain. A well-known problem from this class is writing. The tip of the pencil moves along a surface producing a pressure force upon it. This task aptly illustrates all the characteristics of the matter being discussed. For this reason, instead of using the long expression "motion subject to surface-type constraint" we simply say "*writing task*". In this section, the theoretical specificity of the task will be considered first and then two numerical examples will be presented in detail.

2.1 Theoretical Consideration

If the tip of the end-effector is marked by T and has the Cartesian coordinates (x, y, z), then the surface-type constraint can be expressed by the equality

$$f(x, y, z, t) = 0 \text{ or } f(x, y, z) = 0, \tag{3.10}$$

where the first relation holds for nonstationary constraint (surface that moves or changes its shape), and the latter one for the stationary case. However, this approach to the definition of constraint is not always suitable as it does not support some common cases of mobile surface (Fig. 3.3). For this reason we use the approach that utilizes the functional coordinates s, explained in Section 1.1. Two sets of s-coordinates that suit the surface-type constraint are shown in Figs 3.1a and 3.4. The set from Fig. 3.4 is particularly appropriate if writing is performed with the pencil directed orthogonally. Free coordinates form the subvector $s^f = [s_1 \ldots s_5]^T$. The sixth coordinate alone forms the subvector of constrained coordinates, $s^c = [s_6]$. The constraint implies: $s^c = s_6 = 0$. The reaction force has one scalar component F and it acts along the direction of s_6 as shown in Figs. 3.2a and 3.4.

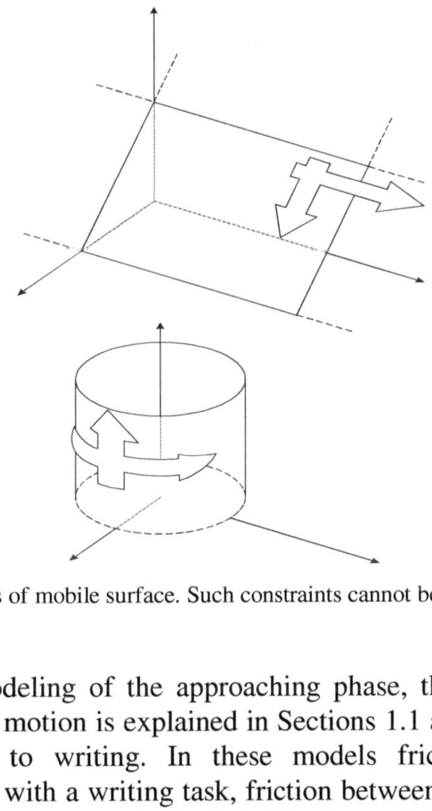

Figure 3.3. Two examples of mobile surface. Such constraints cannot be defined by scalar equality conditions

The dynamic modeling of the approaching phase, the impact, and the phase of constrained motion is explained in Sections 1.1 and 1.2, and can be completely applied to writing. In these models friction effects were neglected. However, with a writing task, friction between the pencil and the surface is important and should be included in the model. Friction force is a vector having the direction opposite to the relative velocity of the tip (with respect to the surface). This means that it lies in the plane tangential to the surface and, thus, has the components along the directions of functional coordinates s_1 and s_2. Let F_{fr} be the 2×1 vector containing these two components. The dynamic model (3.6) should now be supplemented with the friction effect:

$$H(q)\ddot{q} + h(q,\dot{q}) = \tau + J^{c^T}(q)F + J^{(1,2)^T}(q)F_{fr}, \tag{3.11}$$

where $J^{(1,2)}$ denotes the first two rows of the Jacobian (the dimension is 2×6). Note that J^c reduces to the last (the sixth) row of the Jacobian (*dim*=1×6). This relation holds for a stationary constraint, and in a nonstationary case the Jacobian depends on time explicitly: $J^c(q,t)$ and $J^{(1,2)}(q,t)$, as explained in Section 1.1.

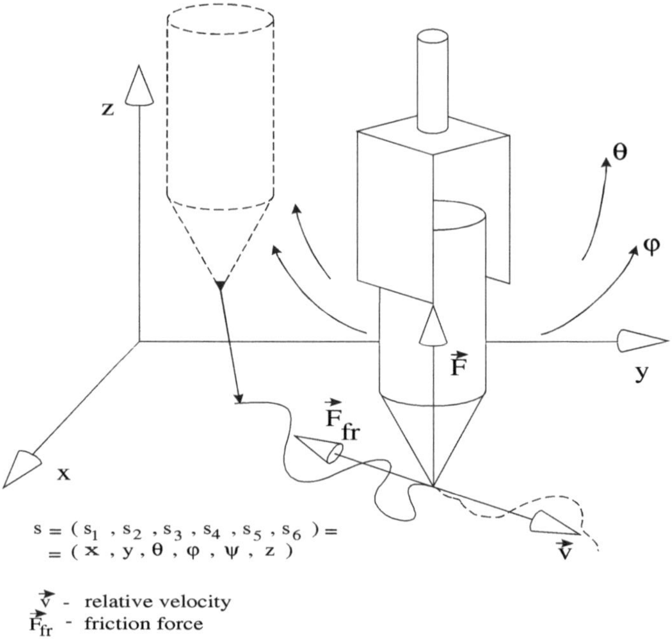

Figure 3.4. The writing task: selection of functional coordinates, the reaction force, and the friction force

The value of the friction force is $|F_{fr}|$. While the tip moves along the surface the magnitude of the friction force may be expressed in the form

$$|F_{fr}| = \mu F, \tag{3.12}$$

where μ is the friction coefficient. The coefficient is usually considered as a constant. However, more exact modeling should introduce the dependence on relative velocity, as shown in Fig. 3.5.

The position of the tip relative to the surface is described by the coordinates s_1 and s_2. Accordingly, the relative velocity involves the derivatives $\dot{s}_1$ and $\dot{s}_2$. In the case of immobile surface (stationary constraint) the derivatives can be expressed in terms of joint velocities:

$$\dot{s}^{(1,2)} = \begin{bmatrix} \dot{s}_1 \\ \dot{s}_2 \end{bmatrix} = J^{(1,2)}(q)\,\dot{q}\,, \tag{3.13}$$

while with the mobile surface (nonstationary case) the expression has to take account of the explicit dependence on time *t:*

$$\dot{s}^{(1,2)} = J^{(1,2)}(q,t)\,\dot{q} + \partial s^{(1,2)}(q,t)/\partial t\,.$$

In the text to follow we mainly concentrate on the stationary problem.

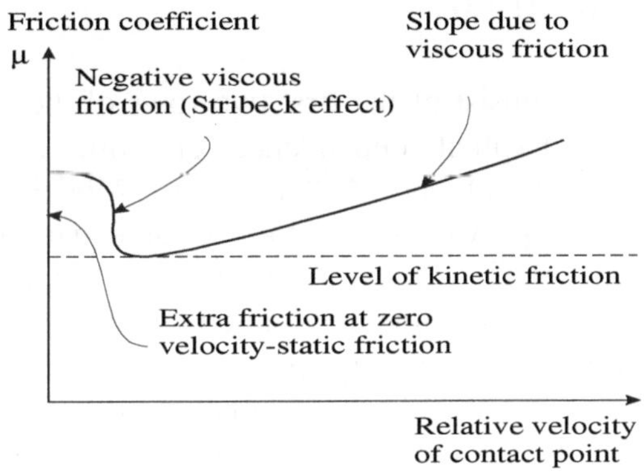

Figure 3.5. Friction depending on relative velocity

The unit vector of relative velocity (two projections onto the directions of s_1 and s_2) is

$$\dot{s}^{(1,2)} \Big/ (\dot{s}_1^2 + \dot{s}_2^2)^{\frac{1}{2}}\,.$$

The two-dimensional friction force is opposite to the unit vector, its magnitude being given by (3.12). Accordingly, the force has the form:

$$F_{fr} = -\,\mu F J^{(1,2)} \dot{q} \Big/ (\dot{q}^T J^{(1,2)^T} J^{(1,2)} \dot{q})^{\frac{1}{2}}\,. \tag{3.14}$$

This form enters the model (3.11).

The model (3.11), with the expression (3.14) for friction, and expression (3.7) for the constraint, defines the dynamics of the constrained motion in writing. With F_{fr} substituted, (3.11) consists of 6 scalar equations and the constraint defined by (3.7) adds one scalar equation. Such system of 6+1 equations can be solved for 6 internal accelerations (vector $\ddot{q}$) and the reaction force F. Calculation of accelerations $\ddot{q}$ enables simulation of the system dynamics.

If the relative motion of the tip along the surface stops for any reason, the situation changes. The discussion on the magnitude and the direction of the friction force (equations (3.12) and (3.14)) no longer stands. The vector F_{fr} now has two independent and unknown components. This increases the entire number of unknowns. However, the fact that the tip does not move over the surface introduces two additional scalar conditions: $\dot{s}_1 = 0, \dot{s}_2 = 0$ or $\ddot{s}_1 = 0$, $\ddot{s}_2 = 0$. The latter two conditions can be written in the form

$$J^{(1,2)}(q)\ddot{q} + A^{(1,2)}(q,\dot{q}) = 0, \tag{3.15}$$

where $J^{(1,2)}$ and $A^{(1,2)}$ consist of the first two rows of the Jacobian J and the adjoint vector A. Explicit dependence on time t appears in a nonstationary case ($J^{(1,2)}(q,t)$ and $A^{(1,2)}(q,\dot{q},t)$). Model (3.11) and the relations (3.7) and (3.15) provide for 6+1+2 scalar equations that can be solved for 6 components of $\ddot{q}$, 1 scalar reaction F, and 2 components of the friction F_{fr}. Such a situation is sometimes called "jamming". Calculation of $\ddot{q}$ enables simulation of the system dynamics during jamming. This analysis holds while the calculated value $|F_{fr}|$ is smaller than μF. When $|F_{fr}|$ reaches the level μF, the tip motion starts and modeling switches to the previously given case.

Note that, for simplicity, this whole chapter deals mostly with a six-joints robot, that is $n = 6$ according to the notations from Section 1.1. However, there is no limitation on expanding the theory to robot mechanisms that have more than six DOFs ($n > 6$). Since the end-effector cannot have more than six DOFs, we talk about redundant robots. Thus, a general constrained problem would deal with an n-joint robot and an m-dimensional constraint. The number of DOFs of the end-effector is $6 - m$. Applying this to the writing task ($m = 1$), we come to a robot arm that has n DOFs, while its end-effector has $6 - 1 = 5$ DOFs. The problem that arises does not follow from the constraints imposed, but from the "surplus" of the arm DOFs. This means that the solution of the mathematical model requires additional conditions that would compensate for the redundancy. In the examples that follow we shall discuss both nonredundant and redundant problems.

In the first example we consider a planar problem, the two-link robot in a contact task. The problem is simplified but suitable for numerical simulation and, more importantly, it keeps all the effects that are of main relevance.

The other example is a spatial redundant robot. The idea of discussing redundancy in this context is not just to demonstrate the possibility of expanding the theory. Redundancy offers some interesting possibilities for controlling contact tasks.

2.2 Two-Link Robot in a Planar Task

Theory of planar problem

Let us consider a robot with two revolute DOFs ($n = 2$) engaged in a writing task (Fig. 3.6). The task consists of two phases. The first one represents free motion towards the surface. The tip moves from point A to point B, where it makes contact. In the next phase the tip moves along the horizontal surface producing a pressure force *F* onto it. This is the phase of contact motion. The tip moves in one direction from B to C, then stops and moves back to B. The internal position vector is $q = [q_1 \; q_2]^T$. With the considered planar robot there are two functional coordinates: $s = [x \; z]^T$. Horizontal motion x is free, that is $s^f = x$, while the vertical motion $s^c = z$ is the one that is affected by the constraint. In the approaching phase z is positive and free. The zero-value defines the contact with the surface.

Let us describe the system (robot) and the task more precisely. The robot arm links are equal: lengths $l_j = 0.3m$, masses $m_j = 3\,kg$, and moments of inertia being $J_j = 0.025kgm^2$. Each joint is driven by a DC motor with a gearbox. The drives are equal and have the following parameters: rotor moment of inertia $I_j = 0.000452\,kgm^2$, constants of torque and back e.m.f. $C_{Mj} = 0.061Nm/A$, and $C_{Ej} = 0.061\,Vs/rad$, armature resistance $R_j = 0.272\,\Omega$, gearbox ratio $N_j = 160$, viscous friction and inductivity being neglected. Now, let us describe the task. The arm starts from the resting position A and moves towards B. Point A has the coordinates: $x_A = 0.15$, $z_A = 0.2598$, and point *B*: $x_B = 0.3$, $z_B = 0$. The tip moves along the straight line AB and reaches B in $T_1 = 1\,s$. The trapezoidal velocity profile has the acceleration/deceleration period $0.2T_1$. Thus, the motion is planned so as to reach the support with zero velocity (no-collision contact). At point B, the contact with the support is made and the robot starts moving along it towards point C (length: BC=0.1*m*). The motion lasts $T_2 = 4\,s$, with a trapezoidal velocity profile and acceleration/deceleration period $0.2T_2$. After reaching C, the tip moves back to B under the same conditions. During the motion from B to C and back to B, the robot has to produce a pressure force $F^* = 100\,N$ upon the support. After the instant of contact, the force should rise from 0 to 100 N linearly in $0.1s$, and after that it should be constant. Between the robot tip and the support, friction occurs. The friction coefficient is μ=0.4. Initial error is Δx=1.28mm Δz=-0.93mm.

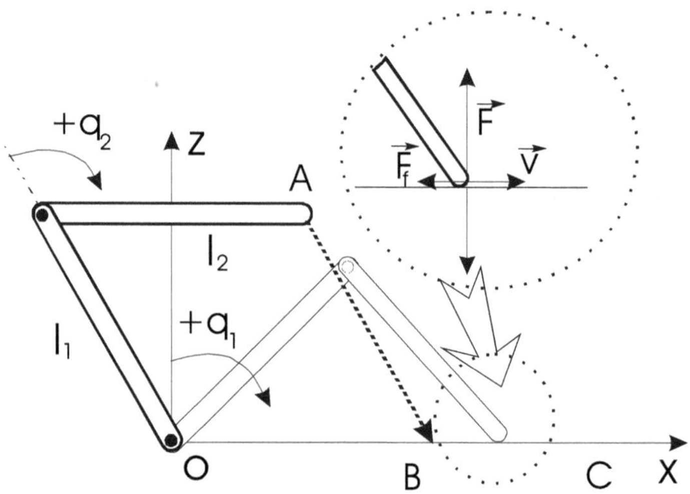

Figure 3.6. Planar robot engaged in a contact task with friction.

In the free motion phase the kinematics of the robot is defined by the model (3.3), (3.4). Since $s^f = x$ and $s^c = z$, the model can be rewritten as:

$$\begin{aligned} \ddot{x} &= J^x \ddot{q} + A^x \\ \ddot{z} &= J^z \ddot{q} + A^z , \end{aligned} \tag{3.16}$$

where J^x, A^x stand instead of J^f, A^f, and J^z, A^z instead of J^c, A^c. For the planar robot the kinematic model is not too complex and the model matrices can be given in analytical form:

$$\begin{aligned} J^x &= \left[l_1 \cos q_1 + l_2 \cos(q_1 + q_2) \quad l_2 \cos(q_1 + q_2) \right], \ \dim = 1 \times 2 \\ A^x &= \left[-l_1 \sin q_1 \dot{q}_1^2 - l_2 \sin(q_1 + q_2)(\dot{q}_1 + \dot{q}_2)^2 \right], \ \dim = 1 \times 1 \\ J^z &= \left[-l_1 \sin q_1 - l_2 \sin(q_1 + q_2) \quad l_2 \sin(q_1 + q_2) \right], \ \dim = 1 \times 2 \\ A^z &= \left[-l_1 \cos q_1 \dot{q}_1^2 - l_2 \cos(q_1 + q_2)(\dot{q}_1 + \dot{q}_2)^2 \right], \ \dim = 1 \times 1. \end{aligned} \tag{3.17}$$

Dynamics of the robot mechanism is described by the model (2.56), that is

$$\begin{aligned} H_{11}\ddot{q}_1 + H_{12}\ddot{q}_2 + h_1 &= \tau_1 \\ H_{21}\ddot{q}_1 + H_{22}\ddot{q}_2 + h_2 &= \tau_2 , \end{aligned} \tag{3.18}$$

where the analytical form of the system coefficients is

$$
\begin{aligned}
H_{11} &= 2J + ml^2(1.5+\cos q_2) \\
H_{12} &= H_{21} = J + 0.5ml^2(0.5+\cos q_2) \\
H_{22} &= J + 0.25ml^2 \\
h_1 &= 0.5ml^2 \sin q_2(-\dot{q}_2^2 - 2\dot{q}_1\dot{q}_2) - 1.5mgl\sin q_1 - \\
&\qquad -0.5mgl\sin(q_1+q_2) \\
h_2 &= 0.5ml^2 \sin q_2\, \dot{q}_1^2 - 0.5mgl\sin(q_1+q_2) \quad ,
\end{aligned}
\tag{3.19}
$$

where $l = l_j$, $m = m_j$, $J = J_j$, and MC is assumed to be at the geometrical center of each link. The complete dynamic model is obtained when (3.18) is combined with the model of actuators (2.75), (2.79) and the transmission (2.81), (2.82). Dynamic model can be solved for the accelerations $\ddot{q} = (\ddot{q}_1, \ddot{q}_2)$.

When z becomes zero, contact with the surface is made, impact arises, and the constrained motion starts. Constrained motion will be discussed first and the impact later. During the constrained motion the constraint is defined by expression (3.7) and it may be rewritten in the form

$$\ddot{z} = J^z(q)\ddot{q} + A^z(q,\dot{q}) = J_1^z\ddot{q}_1 + J_2^z\ddot{q}_2 + A^z = 0\,, \tag{3.20}$$

Simplification (planar model) is reflected on some other expressions such as the ones with friction terms. The dynamic model of constrained motion, equation (3.11), may be rewritten as

$$
\begin{aligned}
H_{11}\ddot{q}_1 + H_{12}\ddot{q}_2 + h_1 &= \tau_1 + J_1^z F + J_1^x F_{fr} \\
H_{21}\ddot{q}_1 + H_{22}\ddot{q}_2 + h_2 &= \tau_2 + J_2^z F + J_2^x F_{fr} \;,
\end{aligned}
\tag{3.21}
$$

where J_i^x and J_i^z are the i-th elements of the corresponding matrices (see expressions (3.17)). The matrix J^x now replaces $J^{(1,2)}$. Friction vector can be expressed in the form (3.14), which reduces to the scalar:

$$F_{fr} = -\mu F\dot{x}/|\dot{x}| = -\mu F \operatorname{sgn}\dot{x}\,. \tag{3.22}$$

The model (3.21), together with the expression (3.20) for constraint, and expression (3.22) for friction, can be solved for three scalar unknowns: the accelerations $\ddot{q}_1$ and $\ddot{q}_2$ and reaction force F.

In the case of jamming, the situation is different. The tip does not move and this is expressed through the condition (3.15), which is simplified to the form:

$$J^x\ddot{q} + A^x = J_1^x\ddot{q}_1 + J_2^x\ddot{q}_2 + A^x = 0. \tag{3.23}$$

Friction force is now unknown (expression (3.22) does not hold), and the condition (3.23) is used with the dynamic model to allow calculation of the friction F_{fr}. Equations (3.21), (3.20), and (3.23) are solved for $\ddot{q}_1$, $\ddot{q}_2$, F, and F_{fr}. The jamming will last until the friction reaches the level μF. When this happens, the tip will start to move and modeling switches again to expression (3.22) for friction force.

Between the approaching phase and the constrained motion phase, impact occurs. Model (3.21) is integrated over the infinitely short impact interval, $\Delta t = [t', t''] \to 0$. One obtains:

$$\begin{aligned} H_{11}\Delta\dot{q}_1 + H_{12}\Delta\dot{q}_2 &= J_1^z F\Delta t + J_1^x F_{fr}\Delta t \\ H_{21}\Delta\dot{q}_1 + H_{22}\Delta\dot{q}_2 &= J_2^z F\Delta t + J_2^x F_{fr}\Delta t \end{aligned} \tag{3.24}$$

Relation (3.20) describing the constraint yields:

$$J_1^z\Delta\dot{q}_1 + J_2^z\Delta\dot{q}_2 = -J_1^z\dot{q}_1' - J_2^z\dot{q}_2'. \tag{3.25}$$

Model (3.24), along with the constraint (3.25), and the friction (3.22), can now be solved for the changes in velocities $(\Delta\dot{q}_1, \Delta\dot{q}_2)$, and the impact impulse $F\Delta t$.

Note that jamming may occur during the impact. In case the tip stops at the point of impact (z stops due to the constraint and x stops due to jamming caused by strong friction). Now, expression (3.22) does not hold and friction becomes unknown. Relation (3.23), expressing the jamming yields:

$$J_1^x\Delta\dot{q}_1 + J_2^x\Delta\dot{q}_2 = -J_1^x\dot{q}_1' - J_2^x\dot{q}_2'. \tag{3.26}$$

Now (3.26) is added to (3.24) and (3.25) to allow the calculation of additional unknown – the friction impulse $F_{fr}\Delta t$.

Simulation

Simulation started from position A and during the approaching phase the robot was controlled by PD regulators. Figure 3.7 presents the results of motion A→B, time histories of the functional coordinates x and z, and the velocities $\dot{x}$ and $\dot{z}$. One may conclude that it was justified to expect a contact with non-zero velocity (error in approaching).

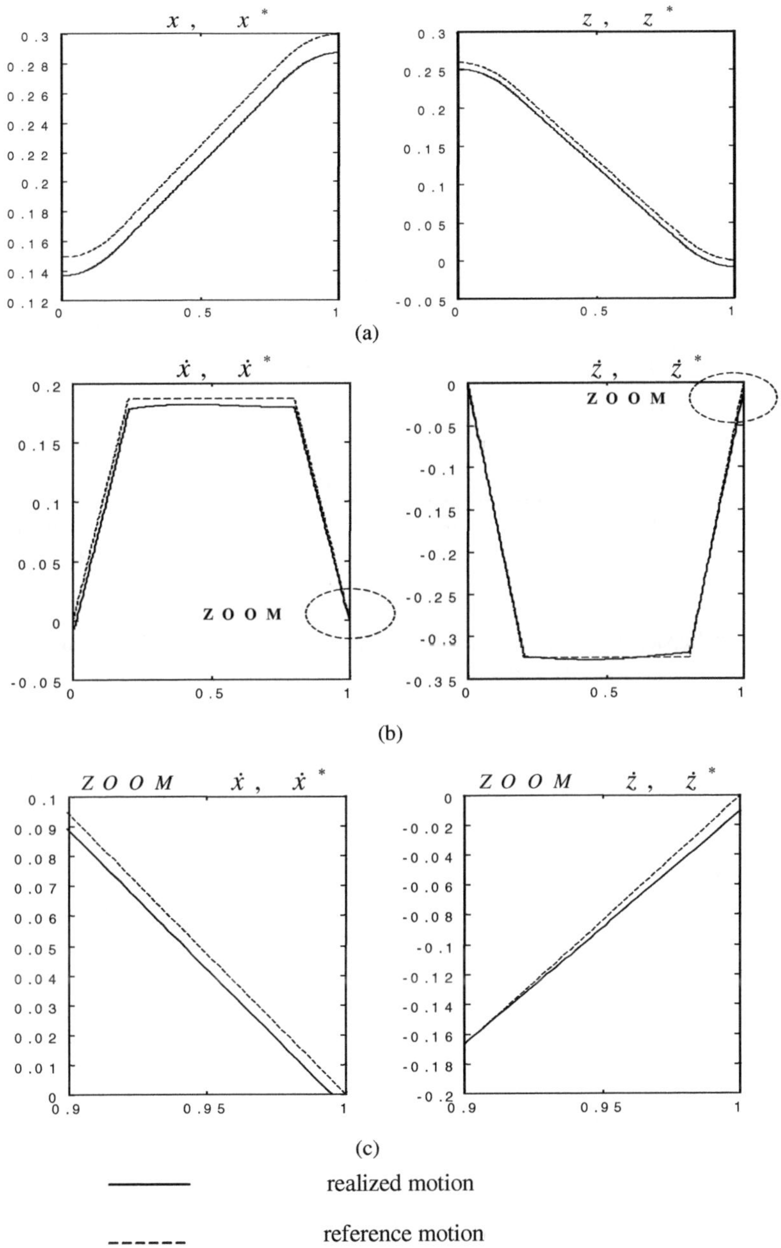

Figure 3.7. Simulation of the approaching phase (A→B):
(a) time histories of functional coordinates x and z (including reference laws), and (b), (c) velocities $\dot{x}$ and $\dot{z}$ (including reference). Zoom view (c) shows that in spite of the no-collision reference trajectory, robot tip reaches the surface with a velocity different from zero.

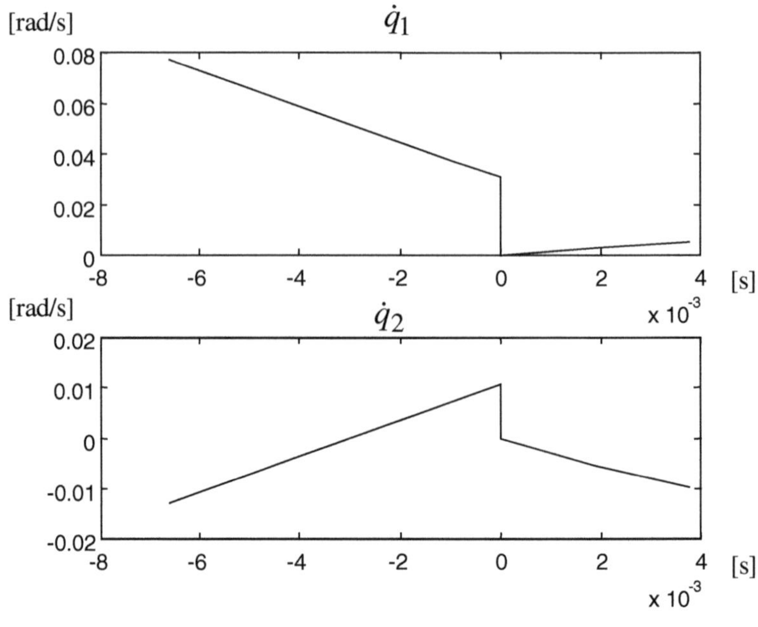

Figure 3.8. Impact. The time histories of joint velocities show that sudden changes appear at the instant of impact. Impact instant is marked by 0.

The impact is shown in Fig. 3.8, giving the change in joint velocities at the instant of impact. After impact, jamming occurs and velocities drop to zero.

During the constrained motion (B→C→B), the hybrid position/force control was applied (see Ref. [4] and Appendix 1). Fig. 3.9 shows the time history of the x-motion and velocity $\dot{x}$. One may see that the robot tip stops in the position C but does not start to move back immediately (as was required). This is the situation of "jamming" (the term jamming will be used for the stick phase of the contact motion). Jamming period is not long (about 0.05 s). After that, the robot tip moves towards B.

Fig. 3.10 shows the reaction force F and the friction force F_{fr}. At point C, the friction changes its sign but not immediately. During jamming, the friction force rises and when it reaches the value μF the tip starts to move again.

Fig. 3.11 shows the errors in tracking the reference motion $x^*(t)$ and reference pressure force $F^*(t)$. Relatively large position error is present all the time. Large force error appears in the transient phase after the impact and after the jamming.

Finally, Fig. 3.12 presents the control inputs.

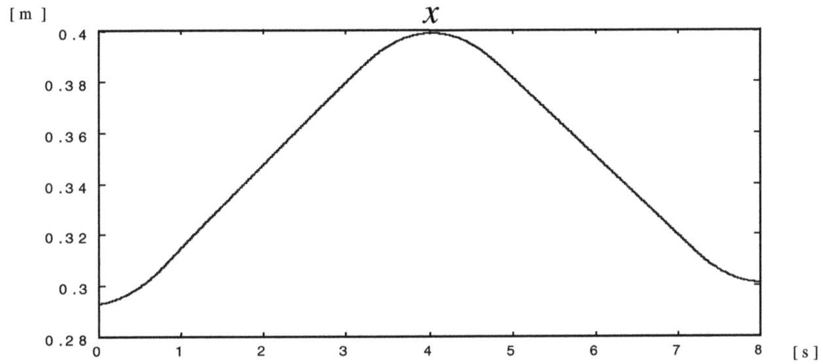

(a)

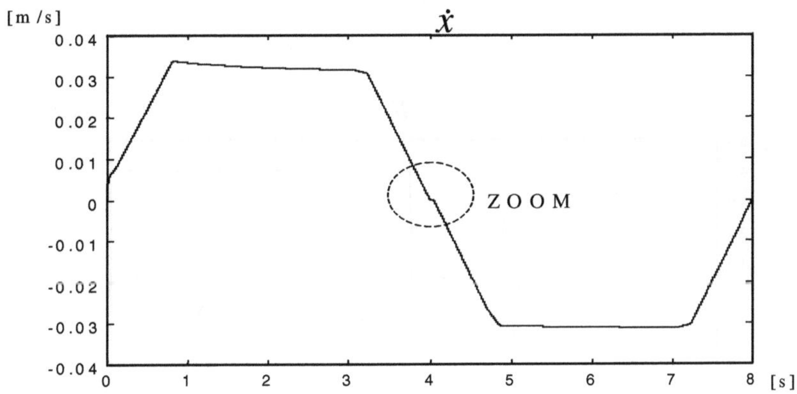

(b)

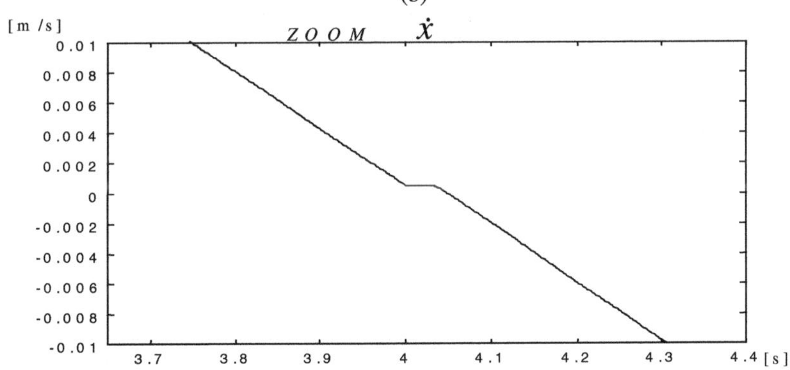

(c)

Figure 3.9. Constrained motion of the tip:
(a) - time history of the coordinate x, and (b), (c) - velocity $\dot{x}$.
Zoom view (c) shows that the tip stays at point C for some time although the reference motion is planned so as to start immediately back at A. This refers to jamming.

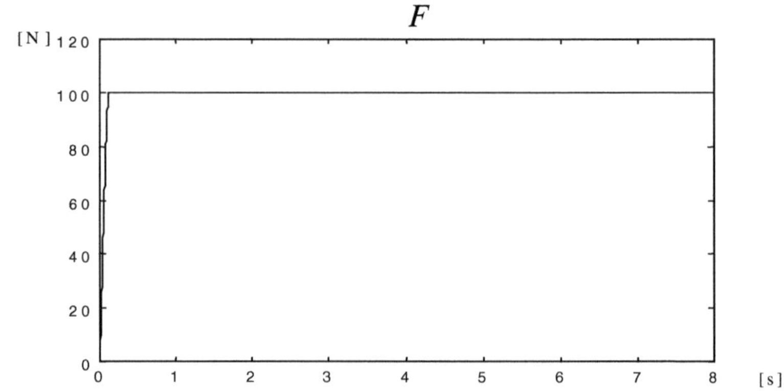

(a)

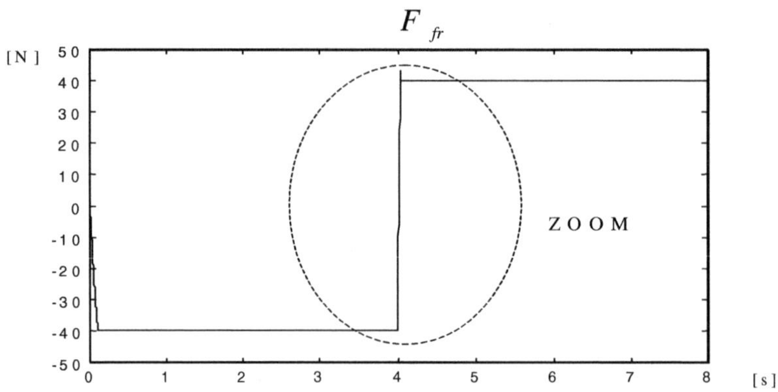

(b)

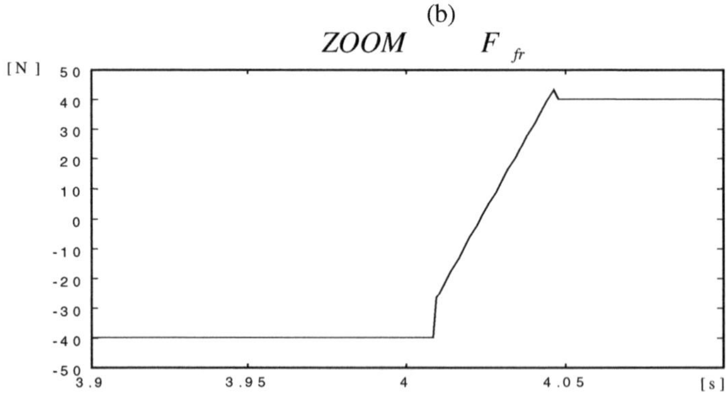

(c)

Figure 3.10. Forces at the contact point:
(a) - reaction force F, and (b), (c) - friction force F_{fr}. Zoom view (c) shows that during jamming friction rises until it reaches the limit μF.

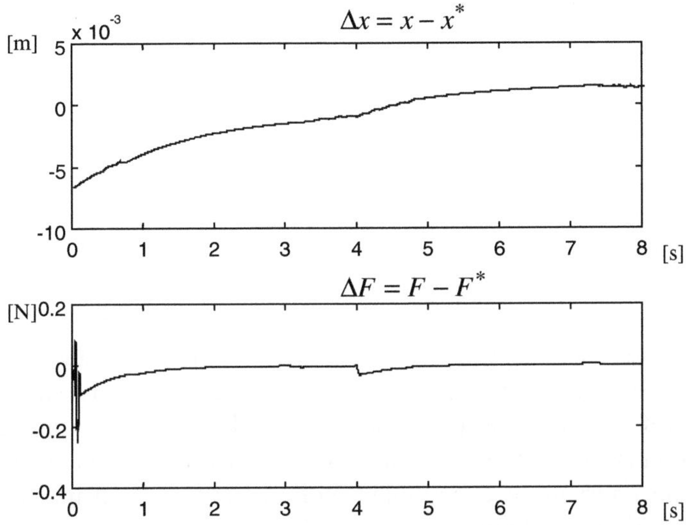

Figure 3.11. Errors: (a) - position error Δx , and (b) - force error ΔF .
Relatively large position error is present all the time. Large force error appears in the transient phase after the impact and after jamming.

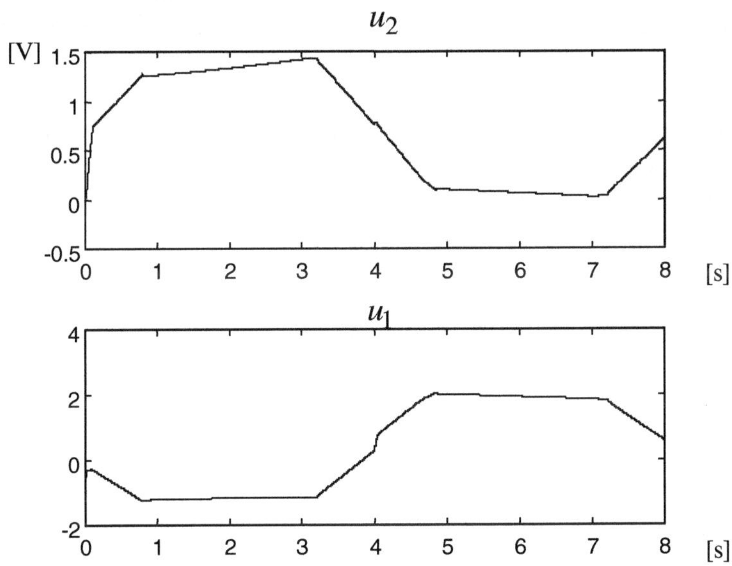

Figure 3.12. Control inputs for DC motors.

2.3 Redundant Robot in a Spatial Task[1]

The spatial task considered in this example consists of writing on a horizontal immobile plane. It is required that the tip writes a series of the Greek capital letter Π. During writing, the pencil should produce a constant pressure force, $F^* = 15$ N, upon the surface. Friction is neglected. Between each two letters, there is an obstacle and the pencil has to jump over it. The task is shown in Fig. 3.13. The required (reference) spatial trajectory is shown in 3.13a, while the projections onto the horizontal (x, y) and the vertical (x, z) plane are shown in 3.13b and 3.13c, respectively. Finally, Fig. 3.13d presents the reference time history of the contact force. Each straight-line segment of the trajectory is performed with a trapezoidal velocity profile, where acceleration and deceleration times are equal and each represents 1/3 of the time used for the uniform motion. Thus, the task includes the intervals of constrained motion (writing) and intervals of free motion (jumping). Although the motion is planned so as to avoid impact, it is expected to appear each time when the pencil touches the surface after performing a jump.

From the kinematic point of view, any writing task could be solved by using a 5-DOF arm. It is capable of producing a tip trajectory on any surface along with keeping the direction of the pencil. Rotation of the pencil about its axis is irrelevant. For the task considered, we use an elbow-scheme arm, shown in Fig. 3.14. Table 3.1 shows its parameters. For actuation, CEM-PARVEX motors F9M4R for joints 1 and 5, and MC17B for joints 2, 3, and 4 were selected, the data being given in Table 3.2. The robot was controlled by using feedforward control based on the dynamic model plus PD feedback at each joint. Different sets of feedback gains were used for the constrained motion (writing) and for the free motion (jumping) in order to achieve most appropriate behavior in each stage of the task.

Simulation was performed by using the dynamic models derived in Chapters 2 and 3. Figures 3.15 and 3.16 show the results of simulation, while Figure 3.15 illustrates the time history of the tip position error, $\Delta = \sqrt{(\Delta x)^2 + (\Delta y)^2 + (\Delta z)^2}$. It can be seen that the error is rather large, close to 5 *mm*. It is considerably larger in the constrained phase of motion. This is due to the fact that the concept of control did not introduce any compensation for the constraint effects. Time history of the contact force is shown in Fig. 3.16. First, we notice that the force has an oscillatory character. The mean-value of the force is very close to the required $F^* = 15$ N. However, oscillations introduced errors up to ± 2.5 N and this is considered unacceptable. At the instant of impact ($t = 3.5$ s) the peak force

[1] This example has been written according to [5].

reaches 23.5*N*. This force might be considered acceptable since it quickly reduces.

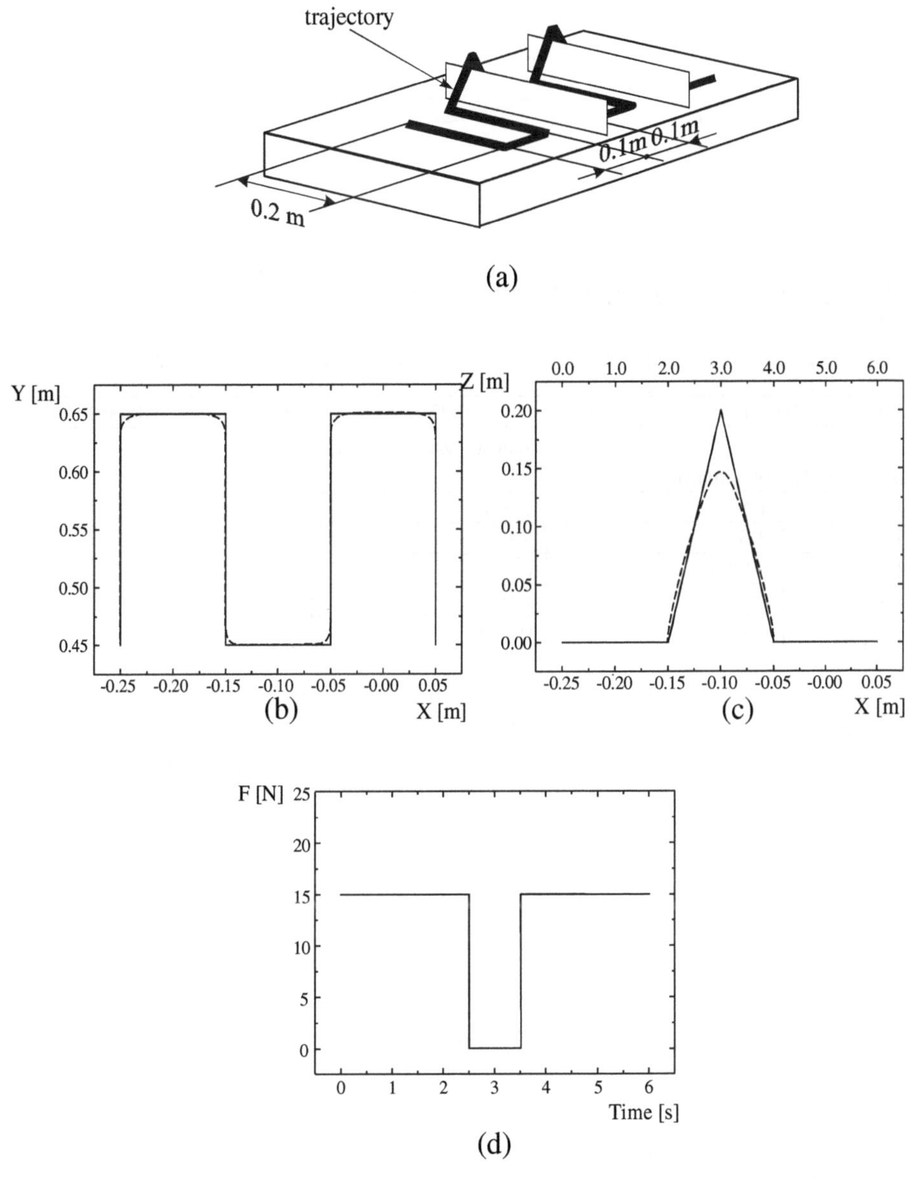

Figure 3.13. Example of writing task. The reference values are shown:
•The required motion of the robot tip is defined by the solid line:
(a) 3D view, (b) *x-y* projection, and (c) *x-z* projection.
•Force required: (d) time history of contact force.

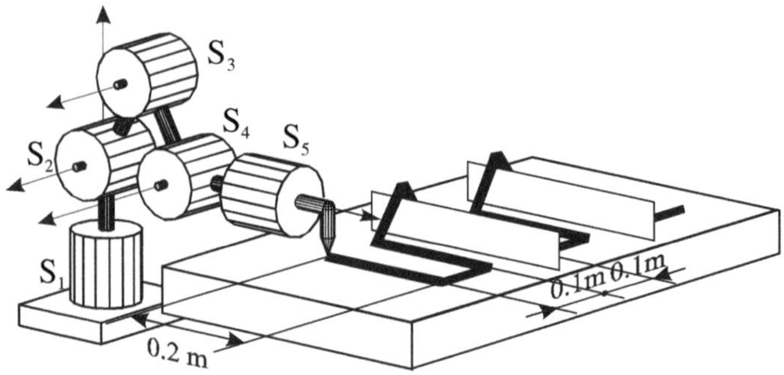

Figure 3.14. Five-DOF robot performing the writing task

Table 3.1. Robot arm data: lengths and inertial parameters

j	1	2	3	4	5
l_j	0.1	0.3	0.3	0.05	0.05
m_j	5	6	6	1	0.3
J_j	0.050	0.270	0.270	0.00250	0.00075

SI metric system is used

Table 3.2. Robot actuators

j	1,5	2,3,4
I_i	$3.5 \cdot 10^{-5}$	$7.9 \cdot 10^{-4}$
C_{Mi}	0.0296	0.0575
C_{Ei}	0.0296	0.0575
R_i	1.1	0.175

The general conclusion is that the five-joint robot, controlled as explained, did not satisfy the performance expected. One way to overcome this situation is to modify the concept of control, e.g., to apply the hybrid control. The other way is to modify the arm by introducing redundancy. In the latter case the control may still be relatively simple and produce good system performance. We apply the redundancy approach as we have already demonstrated the use of hybrid control (see example in 2.2). Besides, our intention is to show the possibilities of dynamic modeling. Redundancy is introduced in the form of three additional joints. Now, the arm consists of the nonredundant part (first five joints, S_1, ... , S_5) and the redundancy (S_6, S_7, and S_8), as shown in Fig. 3.17. Geometrical and dynamic parameters are given in Table 3.3. Redundancy is driven by the DC motors with the parameters presented in Table 3.4. In this example, the redundancy has a twofold role. First, it should allow control of the contact force and absorb the shock during the impact. The other role is to assist the arm in solving high accelerations. Joint S_6 is mainly intended to control the contact force but it also helps the arm to avoid too high torques when the tip moves up quickly. Joints S_7 and S_8 should assist the arm in solving high accelerations in the horizontal plane and, at the same time, they prevent the impact shock from disturbing the arm. If friction force had been introduced, these two joints would have played an even more important role. The action of redundancy in solving high acceleration follows the concept of the so-called distributed positioning [6, 7]. The original motion is decomposed into two components: smooth global motion, and highly accelerated local motion. When the robot performs motion, massive basic configuration (nonredundant part of the arm) solves the smooth component, while the redundancy is responsible for the remaining (local) motion. The smooth global motion is presented in Fig. 3.13 (by using a dotted line) and also in Fig. 3.17.

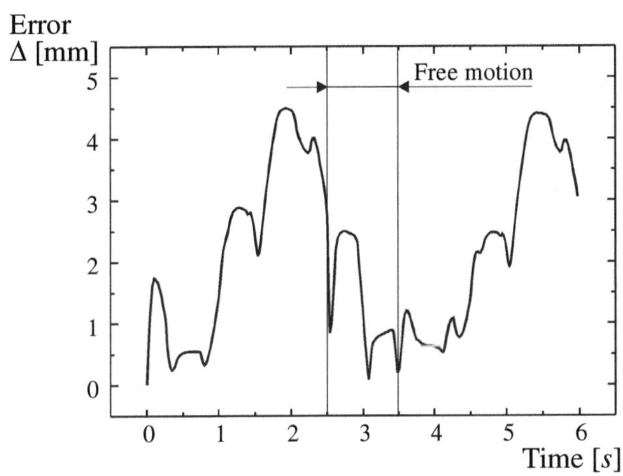

Figure 3.15. Five-DOF robot – tracking error of the robot tip. The error is too large to be acceptable.

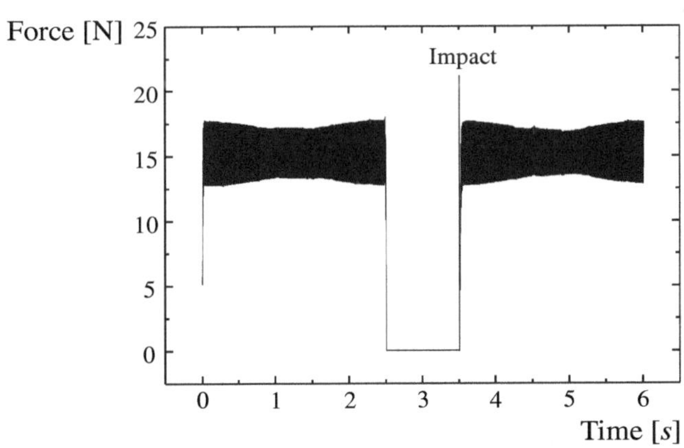

Figure 3.16. Five-DOF robot – time history of the contact force. The mean value is close to the reference force $F^*=15\ N$, but oscillations are too large.

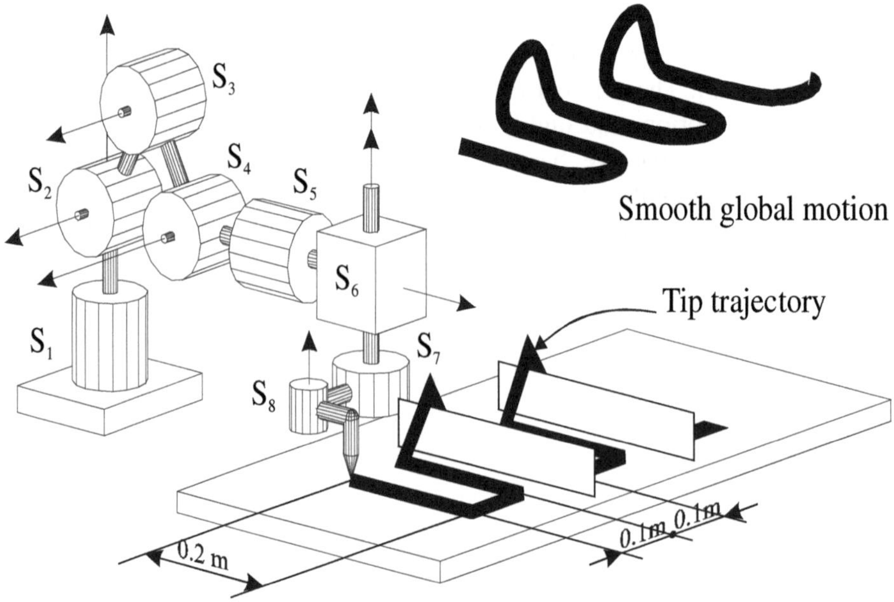

Figure 3.17. Eight DOFs robot (redundant robot) performing the task.
Redundancy is in the form of one linear and two revolute DOFs placed between the basic nonredundant structure and the end-effector.

Table 3.3. Redundancy data: lengths and inertial parameters

j	6,7,8
I_i	$1.2 \cdot 10^{-5}$
C_{Mi}	0.022
C_{Ei}	0.022
R_i	1.38

Table 3.4. Actuators for the redundancy

j	6,7,8
I_i	$1.2 \cdot 10^{-5}$
C_{Mi}	0.022
C_{Ei}	0.022
R_i	1.38

The control applied to the first five joints is the same as previously applied to the nonredundant arm. The problem of controlling the redundancy is more complex. Generally, we did not use feedforward in redundancy. First, we considered the linear joint S_6. During the free motion (jumping over obstacles), PD regulator was used. The gains were selected so as to make the structure "soft", so as to absorb the impact shock. The consequence was the increased tracking error in the phase of jumping, but we did not consider this error relevant (the only request was to avoid the obstacle). For a short impact interval, only suitable D gain was active. In the constrained phase of motion, we applied PI force-feedback regulation. The revolute joints (S_7, S_8) were responsible for the precision. We applied the VSS (variable structure systems) control concept [8].

Simulation was performed by using the dynamic models derived in Chapters 2 and 3. Figures 3.18 and 3.19 show the results of simulation. Figure 3.18 illustrates the time history of the tip position error, $\Delta = \sqrt{(\Delta x)^2 + (\Delta y)^2 + (\Delta z)^2}$. One may conclude that the error is considerably reduced, compared with the nonredundant case. During the constrained motion (writing), maximal error is about 0.07 mm, and this is considered acceptable. During the free motion (passing the obstacle) the error increased due to the control law applied. This caused no problem as the error during jumping is not of interest. Thus, the quality of tracking is considered acceptable. Time history of the contact force is shown in Figure 3.19. The force is close to the prescribed value, $F^*= 15$ N, with the error less than ± 0.1 N. The frequency of force oscillation is high. The integral feedback at joint S_6 annulled the static error. In the instant of impact, there appeared a considerably high force peak (about 22 N) that could not be avoided. It is considered acceptable as the force error decreased quickly. After a transition phase of about 0.01s, the force fell within the bounds, 15 ± 0.1 N.

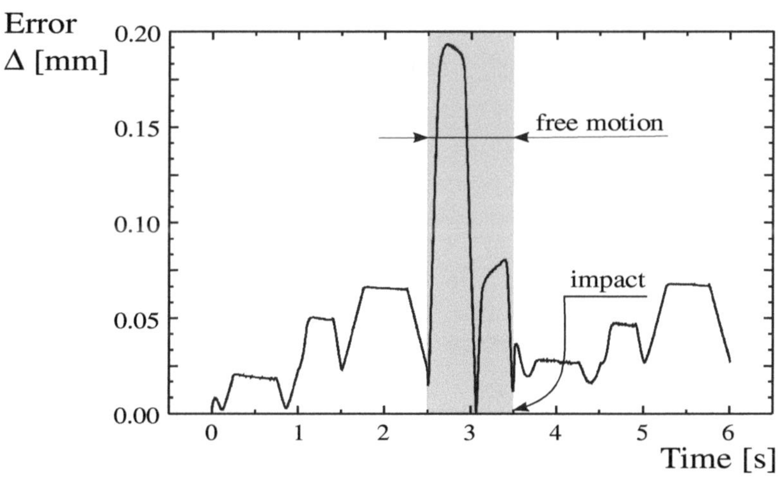

Figure 3.18. Redundant robot (eight DOFs) - tracking error of the robot tip. The error is considerably reduced in the contact phase in which we are primarily interested. Large error during jumping is irrelevant.

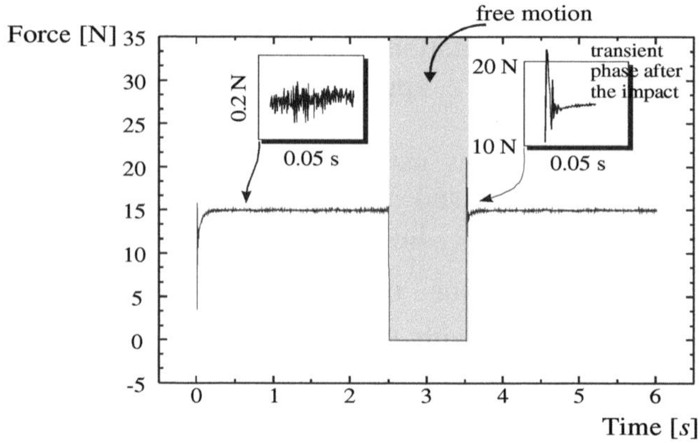

Figure 3.19. Redundant robot (eight DOFs) – time history of the contact force. The error in force is rather small except in the instant of impact. Impact force drops quickly.

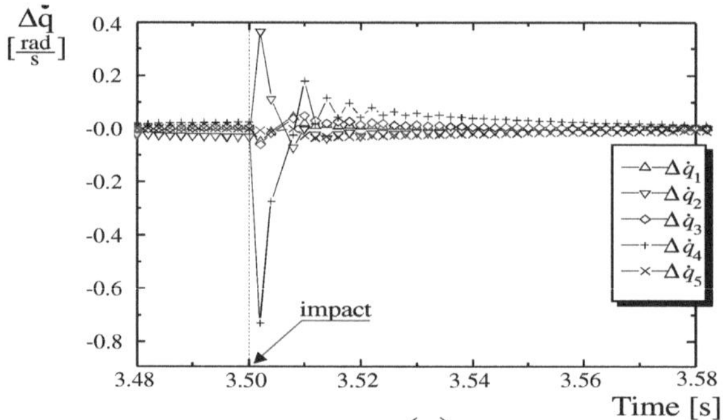

(a)

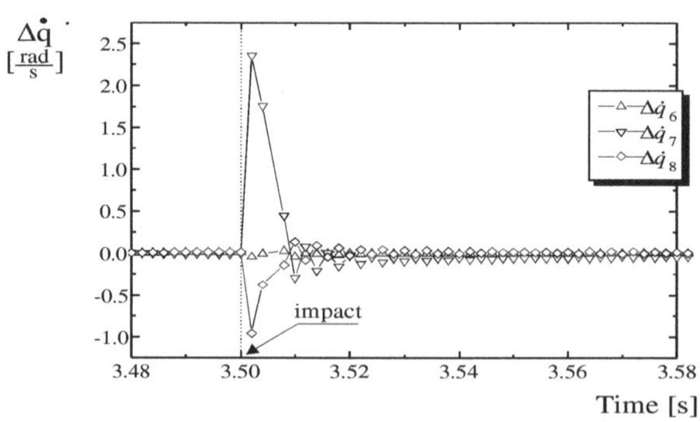

(b)

Figure 3.20. Redundant robot – deviations in joint velocities during the transient phase:
(a)- nonredundant part of the arm, and (b) - redundancy .
Deviations in the redundancy are somewhat larger than those of the nonredundant part. In this way redundancy protects the basic configuration from impact shock. Deviations in velocities diminish quickly.

It is interesting to consider the transient phase after the impact in more detail. Such consideration might support the introduction of redundancy in contact problems. Fig. 3.20 shows the deviation of joint velocities from their reference histories. Fig. 3.20a refers to the basic configuration (nonredundant part of the arm, that is the first five joints), while Fig. 3.20b refers to the redundancy (joints 6–8). At the instant of impact, the deviations in joint velocities of the basic configuration are considerably smaller than those of the redundancy. In this way, the redundancy "protects" the basic configuration. The deviation in the redundant joint velocities are rather large but they diminish quickly. Fig. 3.21 shows the *x-y* trajectory of the tip during the transient phase.

The aim of the presented discussion was not to give a relevant solution to the issue of contact control, but to demonstrate dynamic modeling on a complex example

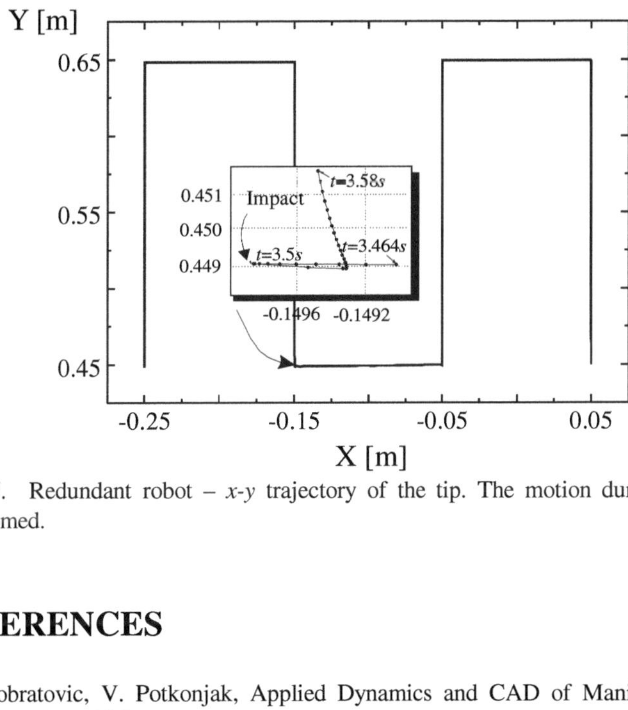

Figure 3.21. Redundant robot – *x-y* trajectory of the tip. The motion during the transient phase is zoomed.

REFERENCES

1. M. Vukobratovic, V. Potkonjak, Applied Dynamics and CAD of Manipulation Robots, Springer-Verlag, 1985.
2. M. Vukobratovic, V. Potkonjak, "Dynamics of Contact Tasks in Robotics. Part I: General Model of Robot Interacting with Environment", Journal of Mechanisms and Machine Theory 34, pp. 923-942, 1999.
3. V. Potkonjak, M. Vukobratovic, V. Matijevic, "Dynamics of Contact Tasks in Robotics. Part II: Case Study in Dynamics of Contact Motion", Journal of Mechanisms and Machine Theory 34, pp. 943-972, 1999.

4. M. H. Raibert, J. J. Craig, "Hybrid Position/Force Control of Manipulators", Journal of Dynamic Systems, Measurements and Control, Vol. 103, pp. 126-133, 1981.
5. G. Djordjevic, Contribution to Dynamics and Control of Redundant Robots, M.Sc. Thesis, Univ. of Nis, Faculty of Electronics, 1992.
6. V. Potkonjak, A. Krstulovic, "Contribution to the Kinematics and Dynamics of Redundant Robots Via Distributed Positioning", Journal of Intelligent and Robotic Systems (Kluwer Academic Publ.) 5, pp. 229-232, 1992.
7. V. Potkonjak, "Applications of Redundant Robots to Constrained-Motion Tasks", Robotica, Vol. 10, pp. 397-407, 1992.
8. V. Potkonjak, G. Djordjevic, C. Milosavljevic, D. Antic, D. Popovic, "Variable Structure Systems for Control of Redundant Robot", Robotics and Autonomous Systems 13, pp. 13-24, 1994.

Chapter 4

SOFT AND ELASTODYNAMIC CONTACTS

In the previous chapter we discussed the contact of the robot arm and its environment by imposing geometrical constraint to the motion of the end-effector. This approach is justified under the following assumptions:

- the end-effector and environment are rigid, i.e. the parts being in contact have such high stiffness that no elastic deformation appears; and

- the environment is immobile or, if mobile, it moves according to a prescribed law that cannot be influenced by the robot.

For these reasons we talked about "rigid contact". However, none of these assumptions are definitely true, but can be applied only as approximations.

In a real situation, there are always some elastic deformations in the contact zone. They appear on both the robot side and the environment side. The influence of deformation on robot dynamics might be significant, depending on the material of the parts being in contact. It is clear that stiffer material allows smaller deformation. However, the magnitude of deformation is not the decisive argument when discussing a contact task. The importance of deformation and the influence on system dynamics have to be considered in the light of the task imposed. If it is required to control the contact force (most common case in contact tasks) then the importance of deformation increases, and its influence on the results of control is rather significant. It is impossible to achieve high quality control without taking into account the effects of elastic deformation. Now, we have to decide on how to model the deformation. In a simplified approach, only spring and damping effects are taken into account. Deformation is regarded as a direct consequence of the contact force and the stiffness and damping of the material. A linear dependence between the force and the magnitude and rate of deformation is usually stated although nonlinear effects may be taken into account also. In this case we talk about *soft contact* (see Section 1). A more complex approach considers deformation as a dynamic process and takes into account the local dynamics of particles (that is, their inertial properties). This approach is called *elastodynamic contact*, and it is discussed in Section 2. In [1], dynamic properties of the environment were introduced first.

If the robot environment is mobile and if the robot can influence this motion, then it is necessary to consider a more complex problem – interaction of the two dynamic systems. As examples of such environment we can mention a positioner, a transport cart, and other devices interacting with the robot. This kind of environment is elaborated in Chapter 5 and is called *complex dynamic environment.*

1. SOFT CONTACT

In this section we consider a simplified approach to modeling of contact tasks. In the contact zone there exists elastic deformation, and we make some assumptions regarding this.

- First, we assume that deformation appears on the environment side only (the robot end-effector is completely rigid). This is justified in many contact problems. For instance, in cutting operations (such as deburring) the cutting tool is much harder than the material being processed. In an assembly task, if a steel peg is to be inserted in a plastic or brass hole, the assumption stands again. However, if the peg and the hole are of the same material, deformation cannot be regarded on one side only (for two-side deformation analysis, see Chapter 5).
- Next, the deformation is considered small. Deviations in different directions are mutually independent.
- Finally, environment is replaced by a spring and a damper along each direction of contact (there is no inertia on the environment side). In this way, contact force depends directly on the magnitude and rate of deformation.

1.1 Mathematical Formulation of the General Problem

We consider a robot arm that has n one-DOF joints. Similarly to the instance considered in Chapter 3, we adopt $n = 6$, assuming that this does not compromise the generality of presentation. Joint coordinates $q = [q_1 \dots q_6]^T$ describe the arm position. The end-effector position is defined by six functional coordinates, $s = [s_1 \ldots s_6]^T$, as explained in Chapter 3. Typical examples are shown in Fig. 3.1. The first phase of a contact task is approaching. It is an unconstrained motion, and thus, all six functional coordinates are free. After the contact is established, the phase of constrained motion starts. The approaching phase was discussed in Chapter 3, and hence, we switch our attention directly to the contact phase. In Chapter 3, rigid constraint was defined by zero values of some functional coordinates: $s^c = 0$, where the subvector s^c contained m constrained coordinates. The remaining $6-m$ coordinates were free and formed the subvector s^f. If the environment is not rigid but deformable (note that the robot end-effector is still considered rigid), the coordinates s^c need not be zero any more. Deviation from zero represents elastic deformation. This means that these coordinates (s^c) are not geometrically constrained. Theoretically, they may take any value. However, the relatively high stiffness keeps the values of coordinates small. Although all s-coordinates are now free, we still keep partition of the vector s into s^f and s^c: $s = [s^{f^T} \; s^{c^T}]^T$. The subvector s^c consists of deformation coordinates while s^f involves those that are "really free". Note that defor-mation coordinates (previously constrained coordinates) may be linear (e.g. $s_6 = z$ in Figs. 3.1a and 4.1) or angular (e.g. $s_5 = \theta$ in Fig. 3.1b). Fig. 4.1 shows the coordinate $s^c = s_6 = z$ in the case of a surface-type constraint. While pressure force acts upon the surface, deformation appears, producing negative values of the coordinate.

During the contact motion, the model of kinematics (relation between the s-coordinates and q-coordinates) is defined by the relations (3.3) and (3.4), that is

$$\ddot{s}^f = J^f(q)\ddot{q} + A^f(q,\dot{q})$$
$$\ddot{s}^c = J^c(q)\ddot{q} + A^c(q,\dot{q})$$

Dynamics of the system is described by the model (3.6), that is

$$H(q)\ddot{q} + h(q,\dot{q}) = \tau + J^{c^T}(q)F\,.$$

Contact force F is now of elastic nature and can be expressed in terms of the stiffness effect and the damping effect. We assume that the deformation is small and so deviations along different coordinates (from set s^c) are mutually independent. For the i-th component of the vector F, it holds that

$$F_i = -k_i s_i^c - b_i \dot{s}_i^c\,, \tag{4.1}$$

where k_i and b_i are the stiffness and the damping coefficients along the direction of the coordinate s_i^c. For all directions from the set s^c one may write a matrix expression

$$F = -ks^c - b\dot{s}^c, \qquad (4.2)$$

where $k = diag[k_i]$ is the m-dimensional stiffness matrix, and $b = diag[b_i]$ is the damping matrix. Note that it was not necessary to adopt pure linear dependence of force in equation (4.1). Any nonlinear effect could easily be added to this expression. With the reaction expressed in the form (4.2), the dynamic model (3.6) can be solved for $n = 6$ unknown accelerations, that is for the vector $\ddot{q}$. Calculation of the accelerations $\ddot{q}$ enables simulation of the system dynamics. Vector of contact forces F may also be calculated. Note that the functional coordinates $(s^c, \dot{s}^c)$ in the relation (4.2) have to be expressed in terms of internal coordinates $(q, \dot{q})$.

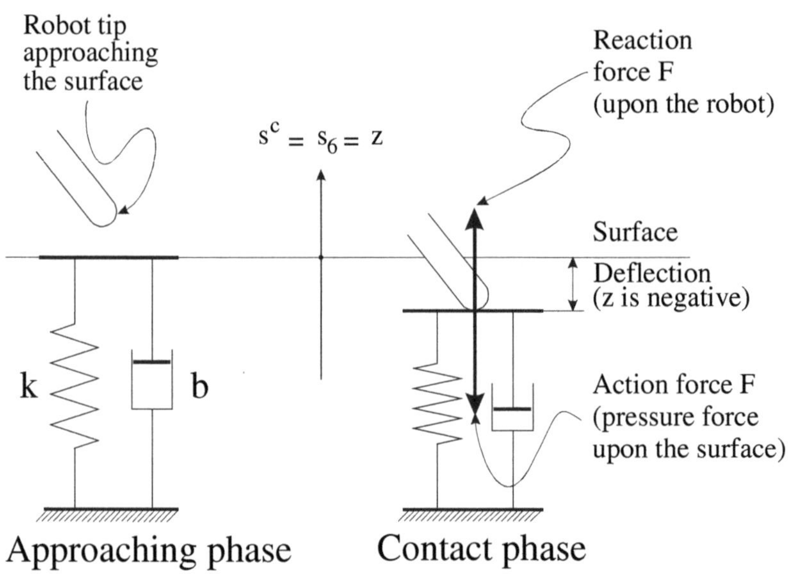

Figure 4.1. Deformation coordinate $(s^c = s_6)$ in the case of a surface-type constraint. Pressure force causes deflection, that is, negative value of the coordinate. Environment is modeled as a spring and a damper.

Let us note another difference from the dynamics of rigid contact. The task starts with approaching – free motion described by means of the model (2.56). s^c is checked constantly and when it becomes zero, the contact is established (we adopt that all components of s^c reach zero values simultaneously (see discussion in Section 1.2. in Ch. 3)). The motion is planned (reference motion) so as to achieve zero-velocity contact. However, the tracking error will produce collision with the constraint. While the rigid-body contact included impact, with the soft contact there is no impact, and the dynamic analysis switches directly to the model (3.6), (4.2). The velocity is continual and the value at the end of the approaching phase is the initial condition for the contact phase. However, the behavior of the system for a certain period after the collision features some specificity. Oscillations in contact force and in deflection appear. With a proper control these oscillations may be quickly decreased. This period is usually called the transient phase. Therefore, we may say that there exists "some kind of impact effect".

1.2 Surface-Type Environment

Theoretical consideration

In this section we discuss the problem of writing over a deformable surface. A similar problem, but involving rigid surface, was elaborated in Section 2. in Ch. 3. In the same manner, we introduce six functional coordinates, vector s, and partition this set into two subvectors: $s^f = [s_1 ... s_5]^T$, containing free coordinates and $s^c = [s_6]$, containing the coordinate that is “conditionally free”. The coordinate $s^c = s_6$ is responsible for the contact with the surface. When it becomes zero, the contact is established and while it is negative, the contact exists. The coordinate represents the deflection. From the standpoint of geometry it is free. However, it is subject to strong elastic force that keeps its value small. Thus, we still use the term “constrained coordinate”.

Kinematics of the contact motion is defined by (3.3) and (3.4). If friction is neglected, the dynamics is described by the model (3.6), along with the expression (4.2). However, friction is a very common effect in writing problems. It can be taken into account in the same manner as was done in the case of rigid surface (see Section 2.1. in Ch. 3). The dynamic model (3.11), that is,

$$H(q)\ddot{q} + h(q,\dot{q}) = \tau + J^{c^T}(q)F + J^{(1,2)^T}(q)F_{fr},$$

is applied instead of (3.6). Expression (4.2) still holds. If the tip moves over the surface, the friction force has the form (3.14), that is

$$F_{fr} = -\mu F J^{(1,2)}\dot{q} \Big/ (\dot{q}^T J^{(1,2)^T} J^{(1,2)}\dot{q})^{\frac{1}{2}}.$$

With this expression for the friction (and (4.2) for the contact force), the model (3.11) can be solved for the accelerations $\ddot{q}$.

If the motion stops for any reason, the two-component friction vector F_{fr} becomes unknown. The additional equation (3.15), that is

$$J^{(1,2)}(q)\ddot{q} + A^{(1,2)}(q,\dot{q}) = 0,$$

is needed to express stopping. Model (3.11) is now solved together with (3.15) for the unknown accelerations $\ddot{q}$ and unknown friction force F_{fr}. This is the jamming situation. When the motion stops, it usually cannot start again immediately. A certain period of time is needed to accumulate enough force. The motion will start when the value $|F_{fr}|$ reaches the level μF.

Simulation example – a planar system

We shall consider the planar example that was described in Section 2.2. in Ch. 3. The robot and the task are shown in Fig. 3.6. The internal position is described by means of $q = [q_1\, q_2]^T$, and the functional motion by $s = [x\, z]^T$. Vector s is partitioned in the same manner as was done in Section 2.2. in Ch. 3: $s^f = x$ is the free coordinate, while $s^c = z$ is the constrained one.

In the approaching phase, the kinematics of planar robot reduces to the relation (3.16), and the dynamics to equation (3.18). The actuators and transmission are modeled applying (2.75) and (2.79) for the DC drives, and (2.81), (2.82) for the gearbox ratio. All the simulation results obtained in Section 2.2. in Ch. 3 for the approaching phase still hold (Fig. 3.7).

When z becomes zero, contact is made, and the phase of constrained motion starts. The final state of the approaching phase represents the initial state for the constrained motion. This phase differs from Example 3.2.2. This time the environment is considered deformable and deflection will appear. Kinematics is described by (3.16), that means both the coordinates x and z are free. While x is "really free", z is "conditionally free". Deflection z is subject to the contact force F that keeps the value of z small. That is why we still use the term “constrained coordinate”. Dynamics of constrained motion may still be described by the model (3.21) and the friction F_{fr} by expression (3.22). The contact force F depends directly on the extent and the rate of deformation. Relation (4.2) that expresses the force now reduces to

$$F = -kz - b\dot{z}. \tag{4.3}$$

We adopt the values for the stiffness and damping: $k = 10^5\, N/m$ and $b = 1\ N/m/s$. The model (3.21), along with (3.22) for the friction, and (4.3) for the reaction, can be solved for the unknown accelerations $\ddot{q}_1$ and $\ddot{q}_2$. After solving the motion (q and then s), the forces F and F_{fr} are computed.

If the tip stops for some reason, it cannot move again immediately. It takes some time to accumulate force and overcome friction. In such jamming situation, friction is an unknown force (expression (3.22) does not hold). Kinematically, jamming is expressed by (3.23). Now, the model (3.21) is combined with (3.23) to solve the three unknowns: the accelerations $\ddot{q}_1$ and $\ddot{q}_2$, and the friction F_{fr}. The reaction force F is still given by (4.3).

Since simulation of the approaching phase (A→B) coincides with that from Section 2.2 in Ch. 3, we start immediately with the results obtained for the contact motion B→C→B. Simulation was performed applying the hybrid position/force control (see Ref. [2] and Appendix 1). Fig. 4.2 presents the motion of the robot tip. Motion in the horizontal direction (coordinate x) is presented in Fig. 4.2a, and vertical motion (z) in Fig. 4.2b. The corresponding velocities ($\dot{x}$ and $\dot{z}$) are shown in Fig. 4.2c and Fig. 4.2d. The horizontal motion represents writing, while the vertical represents deformation (deflection of the surface). Zoom view (e) shows jamming at the position C (the tip remains at point C for some time although the reference motion is planned so as to start immediately back at B). What is characteristic of these results? There are oscillations that appear after the contact is established (transient phase). This is a consequence of the non-zero velocity at the contact point. Oscillations vanish in about 0.5 s due to the action of the control system and the natural damping of the environment.

Fig. 4.3 presents time histories of the contact force F and friction force F_{fr}. Oscillations during the transient phase are again present. In Fig. 4.3c, one may observe friction during jamming. Friction rises towards the value μF, but not with a constant rate (as was the case with rigid environment – see Fig. 3.10). Some oscillations are present. When the friction reaches the level μF, the tip continues to move (back at B).

Fig. 4.4 shows tracking errors: Δx – deviation from the reference motion $x^*(t)$, and ΔF – deviation from the reference force F^*. Relatively large position error is present all the time. Large oscillations of contact force appear in the transient phase.

Finally, Fig. 4.5 shows the control inputs.

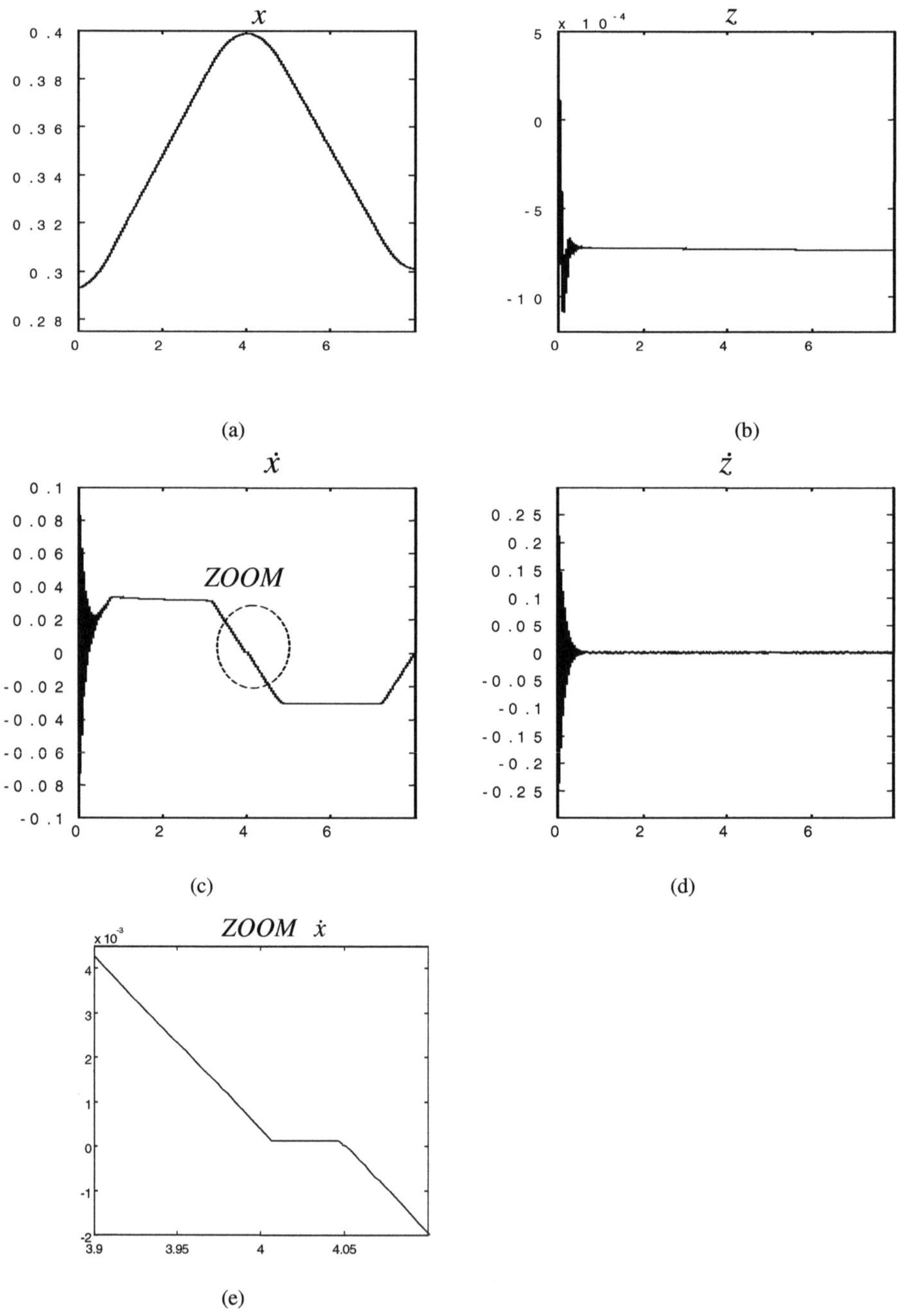

Figure 4.2. Motion of the robot tip during the contact phase:
(a), (b) – Time histories of the horizontal and vertical motion (x and z),
(c), (d) – velocities $\dot{x}$ and $\dot{z}$, (e) – zoom $\dot{x}$.
Oscillations appear during the transient phase and they vanish in about 0.5 s.
Zoom view (e) shows jamming.

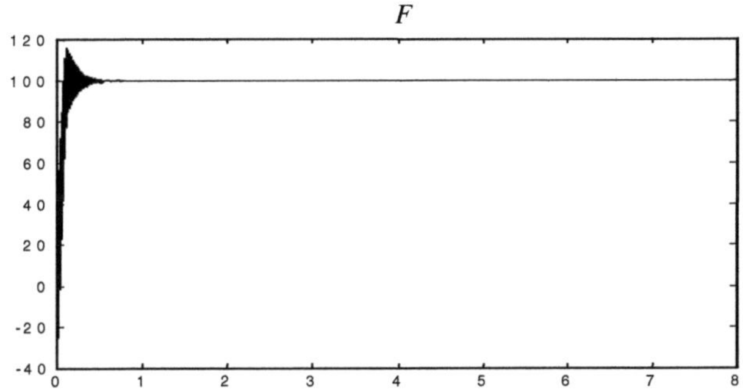

(a)

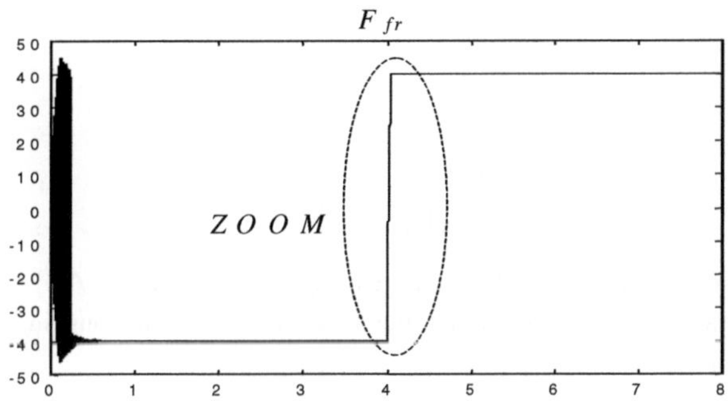

(b)

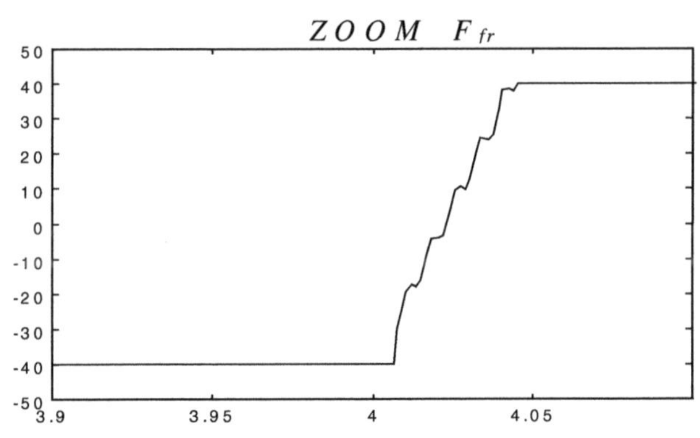

(c)

Figure 4.3. Forces at the contact point:
(a) – contact force F, and (b), (c) – friction force F_{fr}.
Oscillations during transient phase appear with both forces. Zoom view (c) shows some oscillations in friction during jamming: friction rises but not linearly as was the case with the rigid environment (see Fig. 3.10).

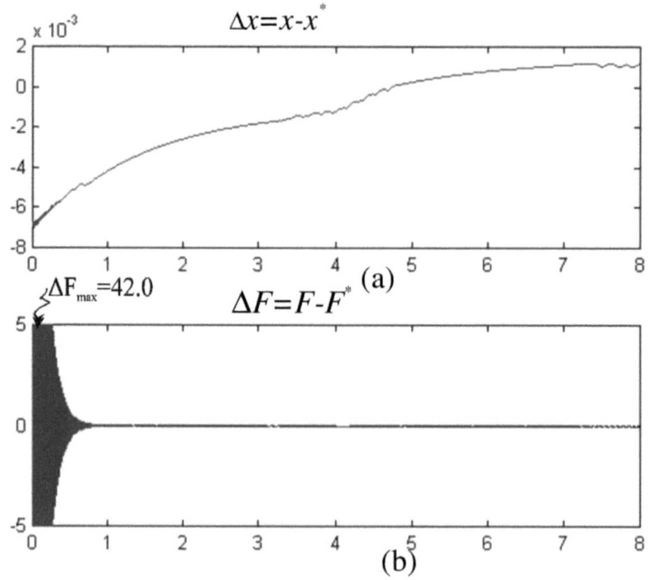

Figure 4.4. Errors: (a) – position error Δx , and (b) – contact force error ΔF . Relatively large position error is present all the time. Large oscillations in contact force appear in the transient phase.

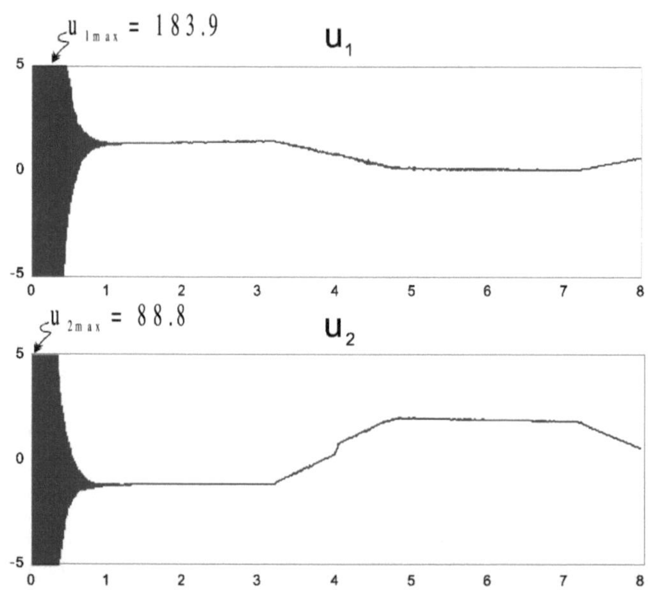

Figure 4.5. Control voltages of DC motors.

2. ELASTODYNAMIC CONTACT

With elastodynamic contact we do not neglect the inertia on the environment side. Since the environment is subject to deformation, the particles exhibit motion. Particles have masses and inertia, and hence, exact modeling requires that we take account of their dynamics. The general approach to elastodynamics would be rather complex, thus we try to make some simplifications that might be justified.

- We assume that deformation appears on the environment side only (the robot's end-effector is completely rigid). This has already been used in Section 1. when soft contact was discussed.
- Deformation is considered small. Deviations in the different directions are mutually independent (also used in Section 1.).

This time, the third assumption from Section 1. does not apply. We do take care of inertia on the environment side.

2.1 Mathematical Formulation of the General Problem

At the beginning, it is necessary to choose some appropriate concept for modeling elastodynamics. We decide to approach the problem via the lumped masses method. This concept applies for each direction of deformation (i.e. along each direction of contact). Let us explain this.

We consider again a robot arm that has n one-DOF joints. Similarly to Section 1.1, we adopt $n = 6$ joint coordinates $q = [q_1 \dots q_6]^T$ to describe the arm position and six functional coordinates $s = [s_1 \dots s_6]^T$ to describe the position of the end-effector (see Fig. 3.1). The discussion of the contact task starts in a way similar to the one given in Section 1.1. The first phase of the task is approaching, an unconstrained motion with all six functional coordinates free (discussed in Chapter 3). After the contact was established and the impact ceased, the phase of constrained motion starts. We discuss this contact phase first. In Chapter 3, the rigid constraint was defined by zero values of some functional coordinates (m of them forming the subvector s^c): $s^c = 0$. The remaining $6 - m$ coordinates were free and formed the subvector s^f. In Section 1., we discussed the environment which was not rigid but deformable: the coordinates s^c were no longer zero. The deviation from zero represented the elastic deformation. The relatively high stiffness kept the magnitudes of the coordinates s^c small. Although all s-coordinates were free, we still kept the partition of the vector s into s^f and s^c: $s = [s^{f^T} s^{c^T}]^T$. The subvector s^c consisted of deformation coordinates, while s^f involved those that were "really free". Fig. 4.1 shows the coordinate $s^c = s_6$ in the case of a surface-type constraint. While the pressure force acted upon the surface, deformation appeared producing negative values of the coordinate. In Section 1., the deformation was considered as a direct consequence of the applied force, and the stiffness and damping characteristics of the material. Now, we use a more complex approach. We take into account the motion and inertia of particles during deformation, and thus, consider deformation as a dynamic process.

Suppose that s_i^c is one of the deformation coordinates from the set s^c. We now discuss the different levels of complexity of the environment deformation model. The discussion is illustrated by the schemes in Fig. 4.6. Case (a) represents the rigid environment where no deformation is possible. This problem was elaborated in Chapter 3. Case (b) considers elastic deformation but with no inertia; such non-dynamic soft contact was discussed in Section 1. Cases (c), (d), (e), and (f) present the dynamic approach; they introduce inertia into the deformation analysis. The simplest model is (c) and it regards the one-stage deformation and one dynamic mode (in the form of an equivalent mass). Sometimes, it is necessary to consider deformation in two stages: external and internal deformation. For instance, if an object has been hardened, the stiffness near the surface is higher than inside the material. The two-stage deformation model is then needed ((d) and (e)). In case (d) one dynamic mode is used, while in case (e) there are two modes (two masses). Generalization may go further, to a multistage deformation and a sequence of equivalent masses (case (f) in Fig. 4.6). Two-stage and multi-stage representations can explain the propagation of deformation through the material of the environment. For the present discussion we selected the problems (c) and (e). (Note that (d) is just a special case of (e)). They illustrate successfully the methodology of modeling. One should note that the coordinate s_i^c may be linear (like z, for example), or angular (like θ). If s_i^c is a linear coordinate (extension and compression), the contact force F (reaction) is measured in N, equivalent mass in kg, stiffness in N/m, and damping coefficient in $N/(m/s)$. For an angular coordinate, the reaction F (now a couple) is measured in Nm, equivalent mass (now the moment of inertia) in kgm^2, stiffness in Nm/rad, and damping coefficient in $Nm/(rad/s)$.

Mathematical modeling of contact motion follows the steps described in the previous section dealing with soft contact. The kinematics is described by relations (3.3), (3.4), and dynamics by the model (3.6). The difference appears in the expression for the contact force F. Let F_i be the i-th component of the vector F (see Fig. 4.7). Relations (4.1) and (4.2) no longer stand.

If the environment is modeled as shown in Fig. 4.7a (one deformation stage and one dynamic mode), Newton's law gives the expression for the force:

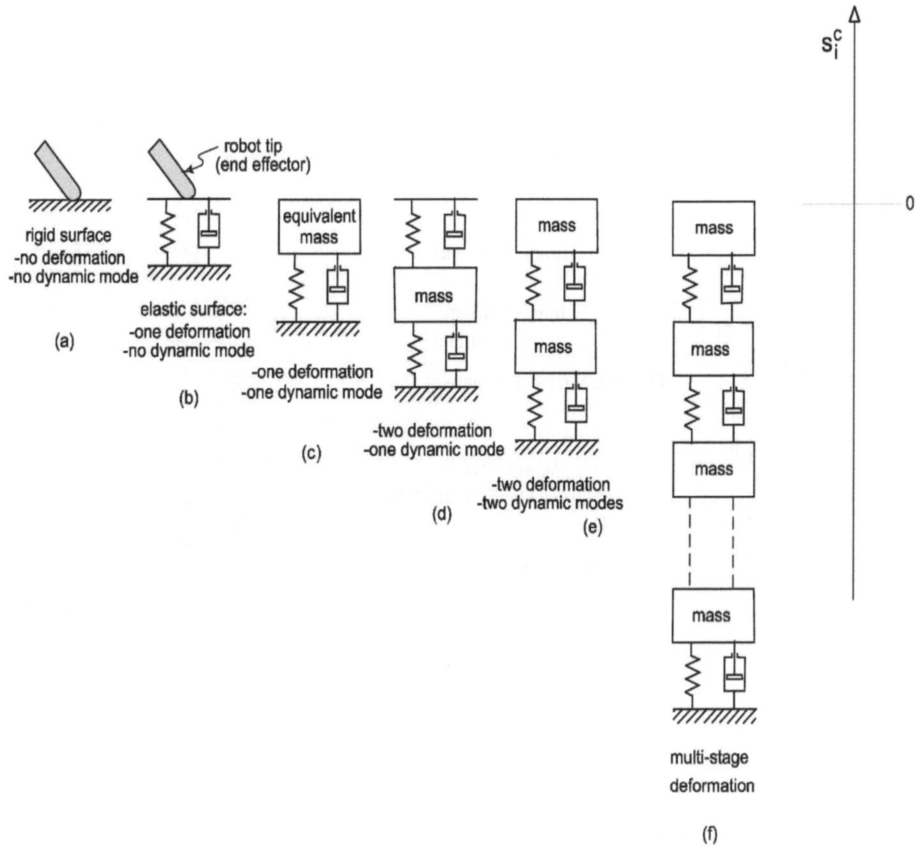

Figure 4.6. Different complexity of the environment model.
One may adopt none, one, or more stages of deformation. This is accompanied by the appropriate number of dynamic modes.

$$F_i = -m_i^{(e)}\ddot{s}_i^c - k_i s_i^c - b_i \dot{s}_i^c, \tag{4.4}$$

where $m_i^{(e)}$ is the equivalent mass, and k_i, b_i are the stiffness and damping along the direction of the coordinate s_i^c. For all the directions from the set s^c (m components) one may write a matrix expression

$$F = -m^{(e)}\ddot{s}^c - ks^c - b\dot{s}^c, \tag{4.5}$$

where $m^{(e)} = diag[m_i^{(e)}]$ is the m-dimensional inertial matrix, $k = diag[k_i]$ is the stiffness matrix, and $b = diag[b_i]$ is the damping matrix. Relation

(4.5) defines the dynamics of the environment. Expression (3.4) is used to relate the coordinates s^c to the internal coordinates q. These two equations are associated to the robot dynamic model (3.6) to allow the calculation of n+2m unknowns: n=6 internal accelerations (that is the vector $\ddot{q}$), m accelerations of the functional motion (vector $\ddot{s}^c$), and m reactions (vector F). Calculation of the accelerations $\ddot{q}$ enables simulation of the system dynamics. Note that it was not necessary to adopt linear spring and damper in equation (4.4). Any nonlinear effect can easily be added to this expression.

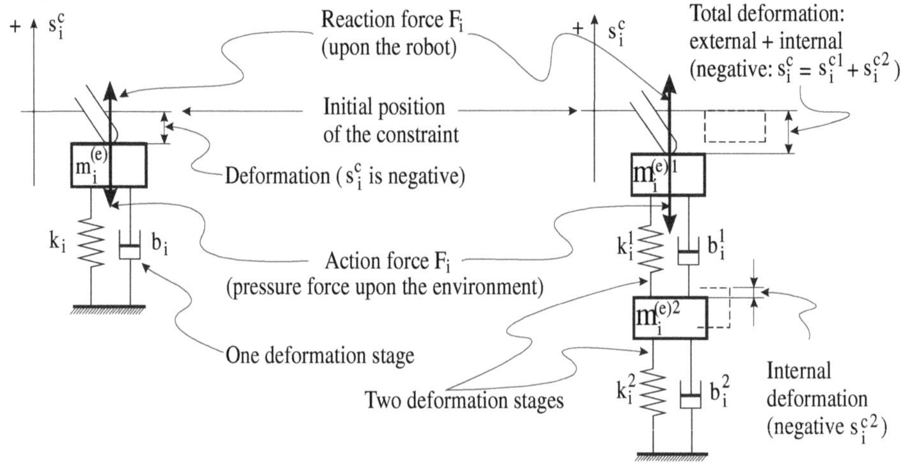

(a) (b)

Figure 4.7. Deformation coordinates and reaction forces in one-stage and two-stage deformation

Let us note another difference of the elastodynamic contact from the soft contact and the rigid-body contact. The task starts with approaching, the free motion towards the constraint. s^c is checked constantly and when it becomes zero, the contact is established (we adopt that all components of s^c reach zero values simultaneously – see discussion in Section 1.2 in Ch. 3). We have already concluded that the end-effector does not reach the constraint with a zero velocity. Thus, the collision occurs. With soft contact, there was no impact and the dynamic analysis switched to the model (3.6), (4.2), describing the contact motion. With the elastodynamic contact, there exists an impact. However, the impact is now different from the impact that appeared with the rigid-body contact. In order to solve the impact of the robot arm and the equivalent mass of the environment, we first integrate the robot model (3.6) over the interval of impact, $\Delta t = [t', t'']$. We assume that the impact is nonelastic and infinitely short. Since $\Delta t \to 0$, the integration of (3.6) over the interval yields

$$H(q')\Delta\dot{q} = J^{c^T}(q')F\Delta t \ , \ \ \Delta\dot{q} = \dot{q}'' - \dot{q}' \tag{4.6}$$

where $\dot{q}' = \dot{q}(t')$ and $\dot{q}'' = \dot{q}(t'')$. Note that the position does not change during the infinitely short impact $(q' = q'')$, and hence, it is not necessary to indicate that H and J^c are calculated for q'. Then, we integrate the model of environment, that is equation (4.5):

$$F\Delta t = -m^{(e)}\Delta\dot{s}^c , \tag{4.7}$$

where $\Delta\dot{s}^c = \dot{s}^c(t'') - \dot{s}^c(t')$. If the equivalent mass was not moving before the impact, then $\dot{s}^c(t') = 0$. Expression (3.4), relating s^c and q, may be written in the first-order form and applied to the instant t'':

$$\dot{s}^c(t'') = J^c\dot{q}(t'') . \tag{4.8}$$

Combination of (4.6), (4.7), and (4.8) leads to the model of impact dynamics in the form:

$$(H + J^{c^T} m^{(e)} J^c)\dot{q}'' = H\dot{q}' + J^{c^T} m^{(e)}\dot{s}^c(t') . \tag{4.9}$$

The model (4.9) can be solved for $\dot{q}''$, that is the internal velocity after the impact. As an additional result, the impact impulse may be calculated using (4.7). The position $q'' = q'$ and the new velocity $\dot{q}''$ represent the initial conditions for the contact motion. One should note that the behavior of the system for a certain period after the collision features some specificity. Oscillations in contact force and in deflection appear, although with proper control, these oscillations may be reduced rapidly. This period is usually called the transient phase. During this period the consequences of the impact can still be observed.

Let us explain briefly how we use the dynamic models to solve numerically the three phases of the contact task. In the approaching phase, the model (2.56) is integrated, constantly checking the value of s^c. When s^c becomes zero, we turn to the impact model (4.9) and solve the change in velocity. We calculate $\dot{q}''$, the velocity after the impact. With the new state, q'' (equal to q') and $\dot{q}''$ (calculated from (4.9)), we start solving the dynamics of the contact motion by integrating the model (3.6), (4.5), (3.4).

If the environment deformation is modeled as shown in Fig. 4.7b (two deformation stages and two dynamic modes), some differences in the dynamic model of contact dynamics will appear. The expression for contact force (4.4), describing the elastodynamics in the direction s_i^c, is now replaced by a more complex dynamic model. Two deformations are introduced: the external s_i^{c1} and the internal s_i^{c2}, and two equivalent masses are involved: $m_i^{(e)1}$ and $m_i^{(e)2}$. The dynamic model consists of two scalar equations:

$$F_i = -m_i^{(e)1}(\ddot{s}_i^{c1} + \ddot{s}_i^{c2}) - k_i^1 s_i^{c1} - b_i^1 \dot{s}_i^{c1}$$
$$k_i^1 s_i^{c1} + b_i^1 \dot{s}_i^{c1} = m_i^{(e)2} \ddot{s}_i^{c2} + k_i^2 s_i^{c2} + b_i^2 \dot{s}_i^{c2} \quad , \tag{4.10}$$

where $k_i^1, b_i^1, k_i^2, b_i^2$ are the coefficients describing the stiffness and damping characteristics. If (4.10) is generalized to cover all directions of deformation $(i = 1,...,m)$ then the matrix expressions are obtained:

$$F = -m^{(e)1}(\ddot{s}^{c1} + \ddot{s}^{c2}) - k^1 s^{c1} - b^1 \dot{s}^{c1}$$
$$k^1 s^{c1} + b^1 \dot{s}^{c1} = m^{(e)2} \ddot{s}^{c2} + k^2 s^{c2} + b^2 \dot{s}^{c2} \quad , \tag{4.11}$$

where F is the *m*-vector of contact forces, s^{c1} and s^{c2} are *m*-vectors of external and internal deformations, $m^{(e)1}$ and $m^{(e)2}$ are the diagonal matrices of equivalent masses, and k^1, b^1, k^2, b^2 are the diagonal matrices of stiffness and damping coefficients. The two matrix equations (4.11) describe the elastodynamics of the environment and replace the model (4.5).

The overall deformation is now

$$s_i^c = s_i^{c1} + s_i^{c2}, \tag{4.12}$$

or in the matrix form:

$$s^c = s^{c1} + s^{c2}. \tag{4.13}$$

This is introduced into the kinematic relation (3.4) yielding:

$$\ddot{s}^{c1} + \ddot{s}^{c2} = J^c(q)\ddot{q} + A^c(q, \dot{q}). \tag{4.14}$$

Relations (4.11) define the dynamics of the environment. Expression (4.14) is used to relate the coordinates s^{c1}, s^{c2} with the internal coordinates q. Now (4.11) and (4.14) are associated to the robot dynamic model (3.6) to allow the calculation of n+3m unknowns: n=6 internal accelerations (that is the vector $\ddot{q}$), 2m accelerations of deformation coordinates (vectors $\ddot{s}^{c1}, \ddot{s}^{c2}$), and m reactions (vector F). Calculation of the accelerations $\ddot{q}$ enables simulation of the system dynamics.

Although the deformation is now treated in a more complex manner, the methodology of solving the impact does not change. Equations (4.6)–(4.9) still hold, with the difference that $m^{(e)1}$ replaces $m^{(e)}$. This is due to the fact that $m^{(e)1}$ is the only equivalent mass directly involved in the impact. However, if a transient phase is considered, then there will be the influence on all the masses ($m^{(e)2}$ in a two-stage model, and $m^{(e)j}$, $j = 2,...$ in a multistage model).

2.2 Surface-Type Environment

Theoretical consideration

The theoretical consideration follows the discussion given in Section 1.2. The only difference is in the expression for the contact force F. Instead of referring to expression (4.2) (that was applied for nondynamic soft contact), we refer to expression (4.5) (for dynamic contact with one-stage deformation), or (4.11) (for two-stage deformation).

In the simulation example we consider again the planar problem that was described in Section 2.2 in Ch 3. The robot and the task are shown in Fig. 3.6. The simulation procedure was described in Section 1.2. The difference in force representation means that the expression (4.3) is replaced by:

$$F = -m^{(e)}\ddot{z} - kz - b\dot{z}, \tag{4.15}$$

in the one-stage deformation modeling, or by

$$\begin{aligned} &F = -m^{(e)1}(\ddot{z}^1 + \ddot{z}^2) - k^1 z^1 - b^1 \dot{z}^1 \\ &k^1 z^1 + b^1 \dot{z}^1 = m^{(e)2}\ddot{z}^2 + k^2 z^2 + b^2 \dot{z}^2 \\ &z = z^1 + z^2 \end{aligned} \tag{4.16}$$

in the case of the two-stage deformation.

When choosing the numerical values for the system parameters, we tried to make them realistic. Although it appeared rather complicated to estimate some parameters (such as equivalent masses), we tried to keep realistic relation between them. With some other parameters (such as stiffness) the problem was much easier. However, we sometimes changed the values of a parameter, giving it a value that might seem less realistic, with the idea of pointing out some dynamic effects. We adopted the following values for the environment characteristics: $m^{(e)} = 1kg$, $k = 10^5 N/m$, $b = 1N/m/s$, for the one-stage model; and $m^{(e)1} = 0.3kg$, $k^1 = 10^6 N/m$, $b^1 = 1N/m/s$,

$m^{(e)2} = 0.7\,kg$, $k^2 = 10^5\,N/m$, $b^2 = 1\,N/m/s$, for the two-stage model. In the two-stage model the external stiffness was adopted to be larger (by the order of magnitude) than the internal one. It is justified by the fact that thermal processing enhanced the hardness of the material surface. However, it might be interesting to examine a completely different situation, i.e. with external stiffness being smaller than the internal one. Why? In a medical application, the robot can be used for massage. In this case, the properties of the human body justify such relation of stiffness parameters. In the text below, when dealing with the simulation of a two-stage environment, we shall also use such estimation of parameters.

Since the simulation of the approaching phase (A→B) coincides with that given in Section 2.2 in Ch. 3, we shift our attention immediately to the results obtained for the impact and the contact motion B→C→B.

Simulation results – one-stage deformation of the environment

We begin simulation by using the one-stage deformation model. Simulation is performed applying three types of control: "classical" hybrid position/force control (Ref. [2] and Appendix 1), dynamic position control (Ref. [3, 4] and Appendix 2), and dynamic force control (Ref. [3–5] and Appendix 2).

Fig. 4.8 shows the changes in internal velocities at the instant of impact.

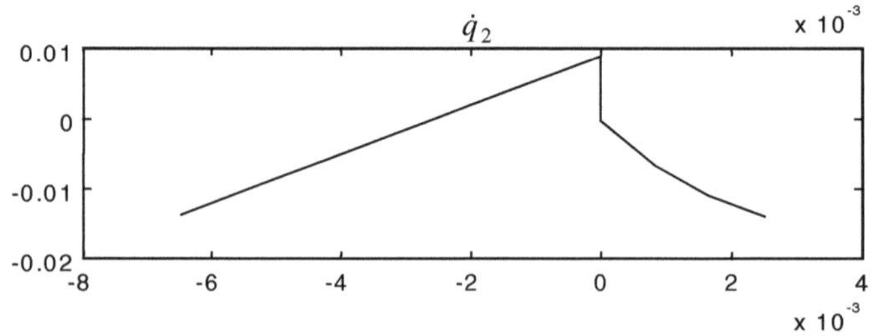

Figure 4.8. Impact. Time histories of joint velocities show that sudden changes appear at the instant of impact. Impact instant is marked by 0.

The set of figures 4.9–4.12, show the results of simulation in the case of the classical hybrid control. Fig. 4.9 presents the motion of the robot tip.

Motion in the horizontal direction (coordinate x) is presented in Fig. 4.9a, and vertical motion (z) in Fig. 4.9b. The corresponding velocities ($\dot{x}$ and $\dot{z}$) are shown in Fig. 4.9c and Fig. 4.9d. Horizontal motion represents writing while the vertical one represents deformation (deflection of the surface). One may observe oscillations during the relatively long transient phase.

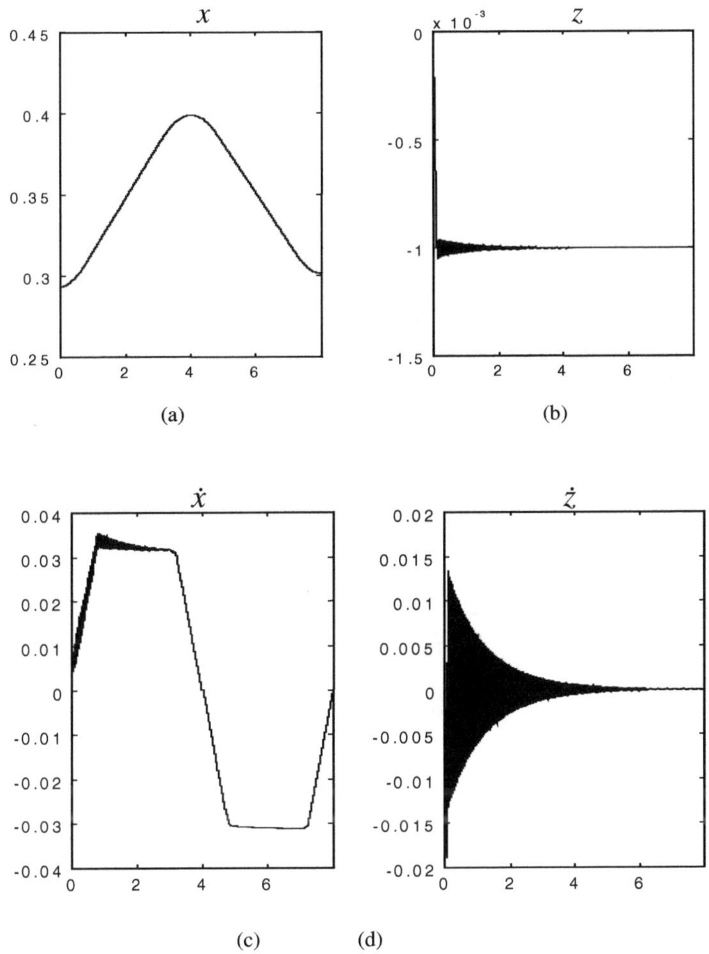

Figure 4.9. Simulation results – classical hybrid control applied.
Motion of the robot tip during the contact phase:
(a), (b) – Time histories of the horizontal and vertical motion (x and z),
(c), (d) – velocities $\dot{x}$ and $\dot{z}$.
One may observe oscillations during a relatively long transient phase.

Fig. 4.10 presents the time histories of the contact force F and friction force F_{fr}. In Fig. 4.10c, one may observe the jamming period. The tip remains in position C for some time. During jamming the friction rises (but not at a constant rate) towards the value μF. When the friction reaches the level μF the tip continues to move (back to B).

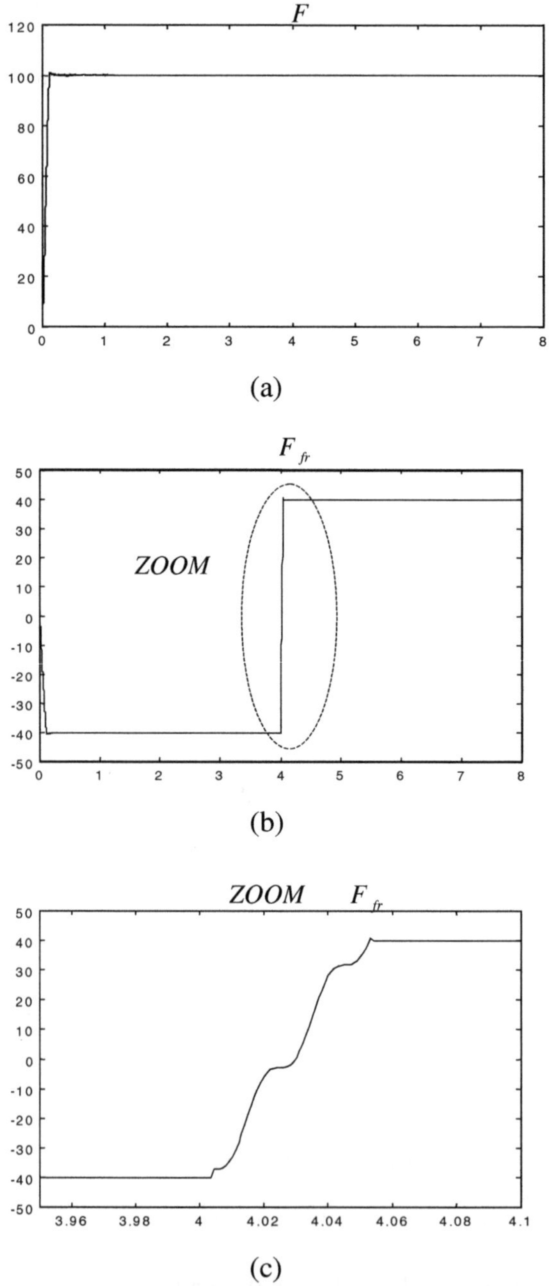

Figure 4.10. Simulation results – classical hybrid control applied.
Forces at the contact point: (a) – contact force F, and (b), (c) – friction force F_{fr}.

Fig. 4.11 presents the tracking errors: Δx– deviation from the reference motion $x^*(t)$, and ΔF– deviation from the reference force F^*. Oscillations in ΔF are present.

Finally, Fig. 4.12 shows the control inputs. Oscillations in control are relatively large and decrease slowly.

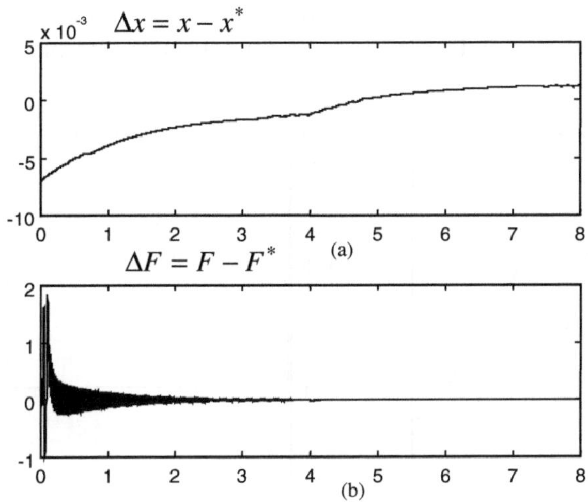

Figure 4.11. Simulation results – classical hybrid control applied.
Errors: (a) – position error Δx, and (b) – contact force error ΔF.
Oscillations in ΔF are present.

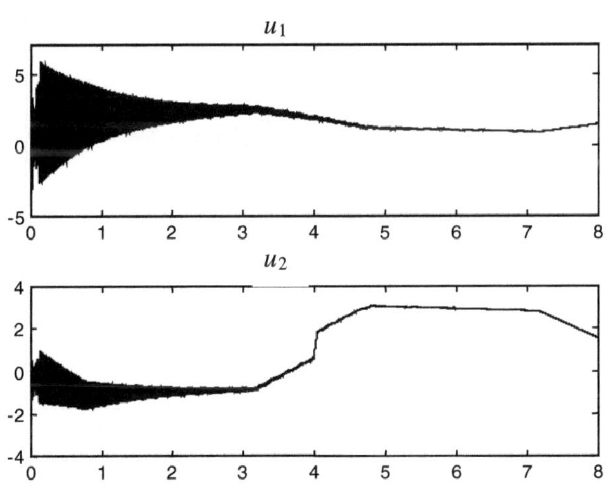

Figure 4.12. Simulation results – classical hybrid control applied.
Control voltages of DC motors.
Oscillations in control are relatively large and decrease slowly.

The set of figures 4.13–4.16 show the results of simulation in the case of the dynamic position control. The same characteristics of motion are considered as in the case of hybrid control. Fig. 4.13 presents the motion of the robot tip (horizontal – writing, and vertical – deflection); Fig. 4.14 presents time histories of contact force F and friction force F_{fr}; Fig. 4.15 presents the tracking errors Δx and ΔF. Finally, Fig. 4.16 shows the control inputs. One may observe that the transient-phase oscillations are considerably reduced and jamming is eliminated. The tracking of the x-motion is strongly improved. This includes elimination of jamming at C (see Fig. 4.14c).

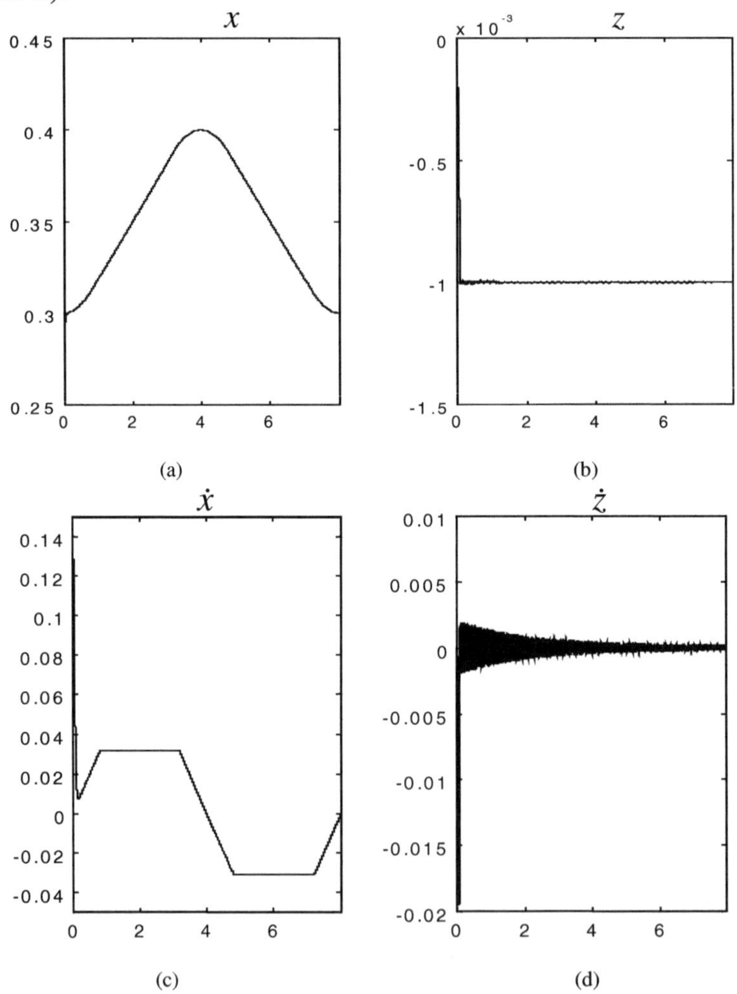

Figure 4.13. Simulation results – dynamic position control applied.
Motion of the robot tip during the contact phase:
(a), (b) – Time histories of the horizontal and vertical motion (x and z),
(c), (d) – velocities $\dot{x}$ and $\dot{z}$.

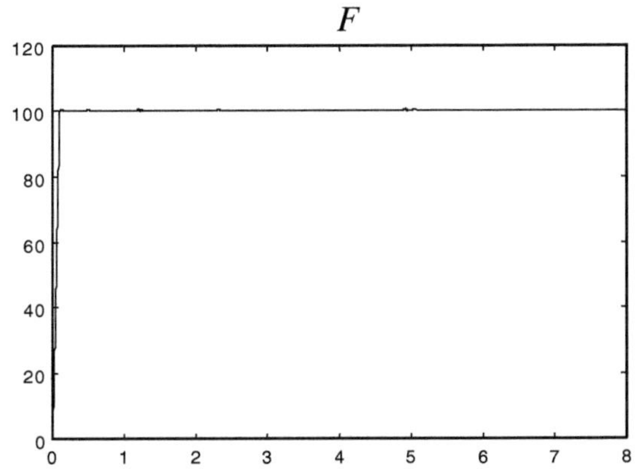

(a)

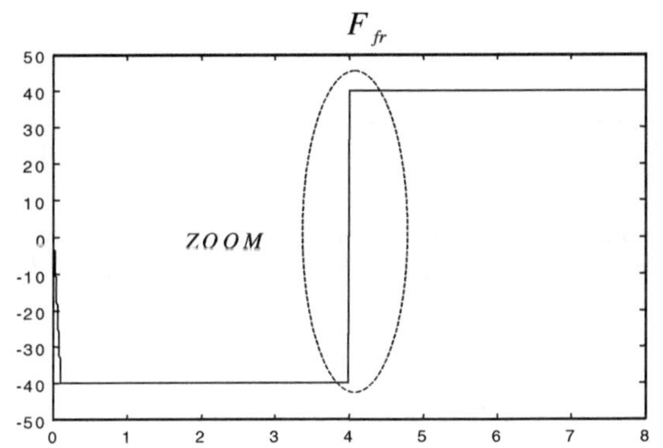

(b)

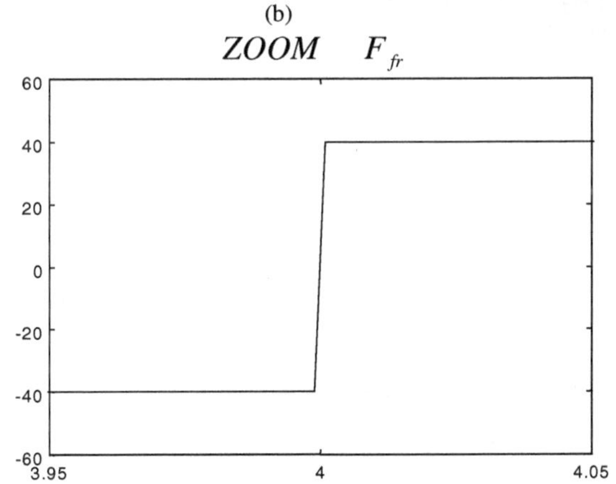

(c)

Figure 4.14. Simulation results – dynamic position control applied.
Forces at the contact point: (a) – contact force F, and (b), (c) – friction force F_{fr}.

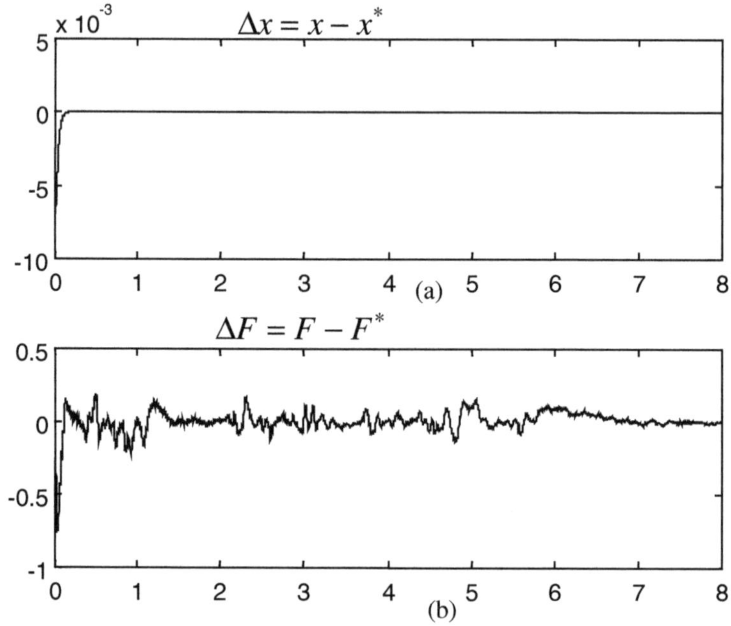

Figure 4.15. Simulation results – dynamic position control applied.
Errors: (a) – position error Δx , and (b) – contact force error ΔF .

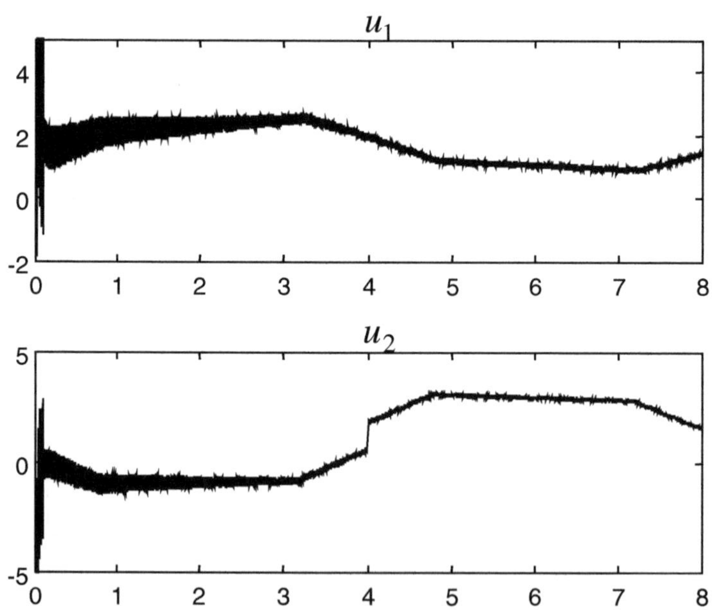

Figure 4.16. Simulation results – dynamic position control applied.
Control voltages of DC motors.

The set of figures 4.17–4.20, show the results of simulation in the case of the dynamic force control. The same characteristics of motion are considered as for the previous two control laws. Fig. 4.17 presents the motion of the robot tip (x and z); Fig. 4.18 presents the time histories of the forces F and F_{fr} ; Fig. 4.19 presents the errors Δx and ΔF , while Fig. 4.20 shows the control inputs. One may observe the improved tracking of the pressure force.

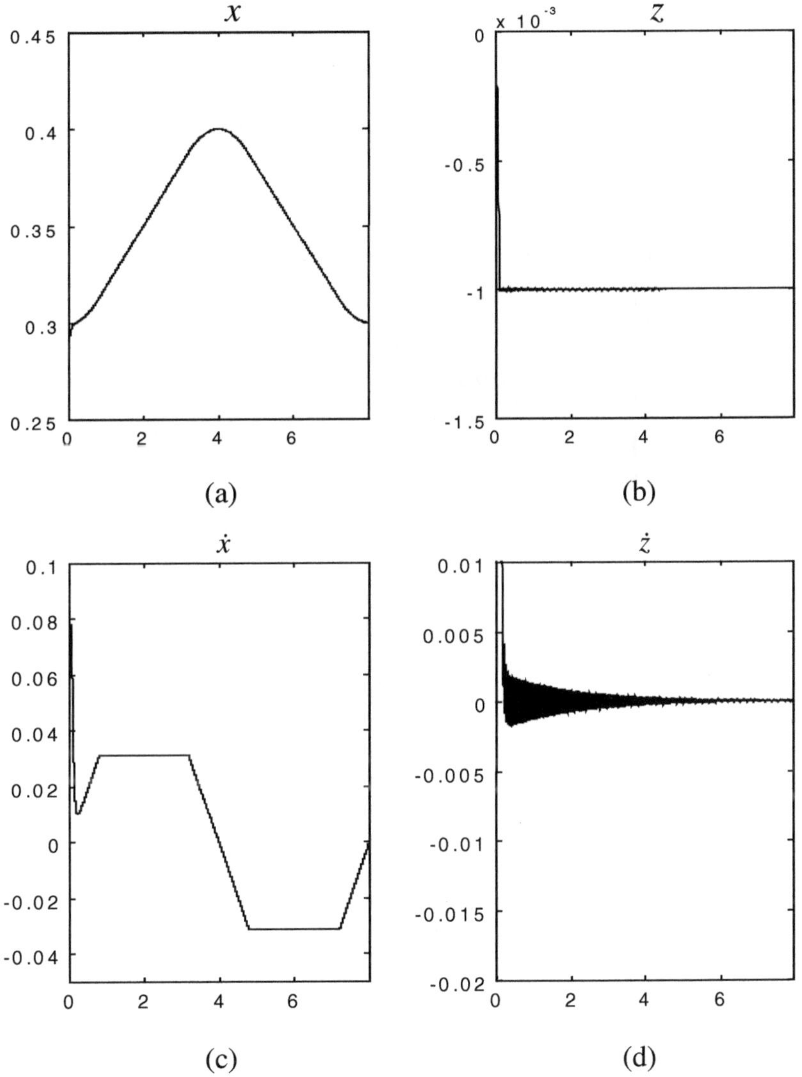

Figure 4.17. Simulation results – dynamic force control applied.
Motion of the robot tip during the contact phase:
(a), (b) – Time histories of the horizontal and vertical motion (x and z),
(c), (d) – velocities $\dot{x}$ and $\dot{z}$.

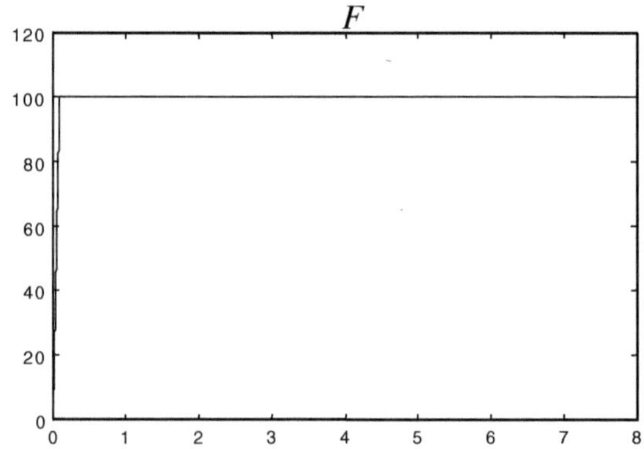

(a)

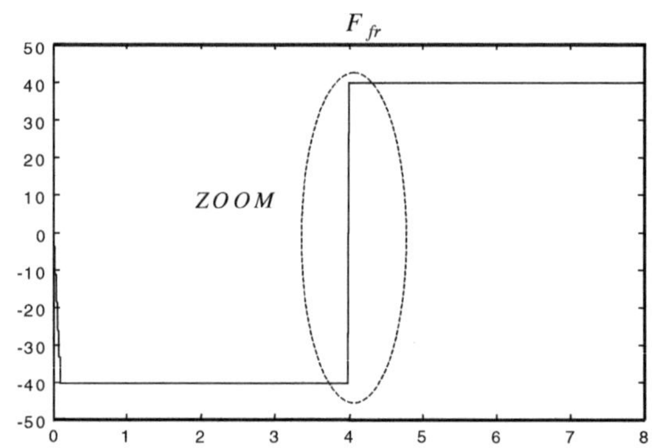

(b)

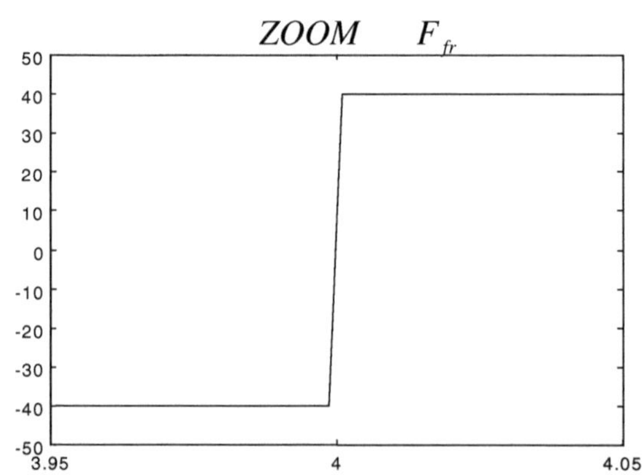

(c)

Figure 4.18. Simulation results – dynamic force control applied.
Forces at the contact point: (a) – contact force F, and (b), (c) – friction force F_{fr}.

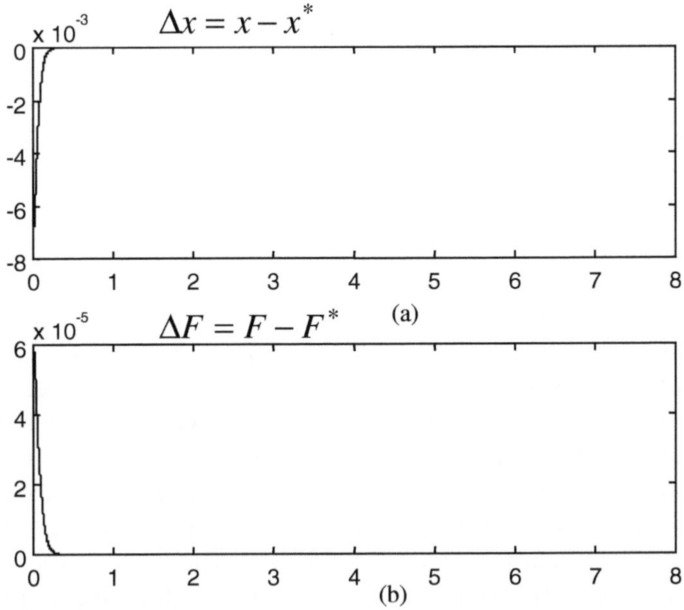

Figure 4.19. Simulation results – dynamic force control applied.
Errors: (a) – position error Δx , and (b) – contact force error ΔF .

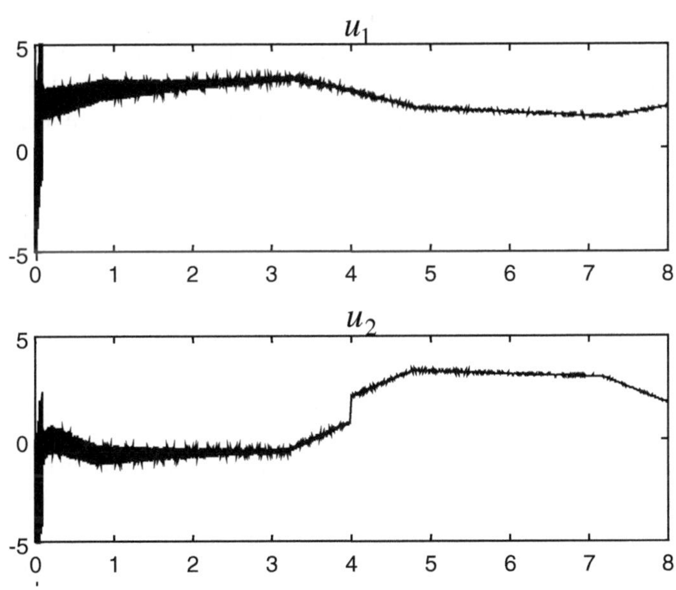

Figure 4.20. Simulation results – dynamic force control applied.
Control voltages of DC motors.

The objective of this study was not to give an elaborated account of control. Various control schemes were used to illustrate the possibilities of modeling and simulation. However, once the results of simulation of different control strategies are presented, it is natural to think about their relation. Without the idea of giving a competent answer concerning the efficiency of the applied control, we shall only compare the results of simulation and draw some conclusions. The conclusions will hold for the example considered, but they may give some guidelines for the future elaboration of control problems in contact tasks.

From the results obtained for classical hybrid control one may find that there exists a considerable error in tracking the x-motion. This is due to the presence of the friction force in this direction, not taken into account by the control. From the standpoint of control, friction is an unmodeled dynamic effect. One may also notice the presence of jamming at position C. Although the motion is planned so as to start back at position B immediately after reaching C, this does not happen. The robot tip remains at position C for some time. This is also caused by friction. The system waits until the position error is sufficient to produce the control along the x-axis that can overcome the friction.

Dynamic position control is a rather different concept. The position is controlled in all directions and through the dynamic model the force is also controlled. In this way, the problem of error along the coordinate x is eliminated and also the problem of jamming. Here, we do not discuss the problems that may appear in the realization of this control concept.

Similar improvements can be obtained if the hybrid control was modified so as to take care of the friction force along the x-axis. However, one should still be careful when selecting the feedback gains. When stabilizing the force in the directions of constrained coordinates, one has to ensure that the position in these directions is also stable. With small damping of the environment, instability of the position may appear.

Simulation results – two-stage deformation of the environment

First, simulation was performed applying the classical hybrid position/ force control.

Fig. 4.21 shows the changes in internal velocities at the instant of impact. The set of figures 4.22–4.25, show the results of simulation in the contact phase. Fig. 4.22 presents the motion of the robot tip: $x-$ writing, $z-$ total deformation, z^1 and z^2- external and internal deformations. It is evident that the external deformation is smaller than the internal one (by one order of magnitude: $0.1mm$ against $1mm$). Fig. 4.23 shows the time histories of the forces F and F_{fr}; Fig. 4.24 gives the errors Δx and ΔF; and Fig. 4.25

shows the control inputs. One may notice several characteristics of the system behavior. There exists a relatively long transient phase with the oscillations in the system variables. Oscillations can be observed in the deformation, tip velocity, force error, and control voltages. One may also observe the jamming interval in Fig. 4.23c.

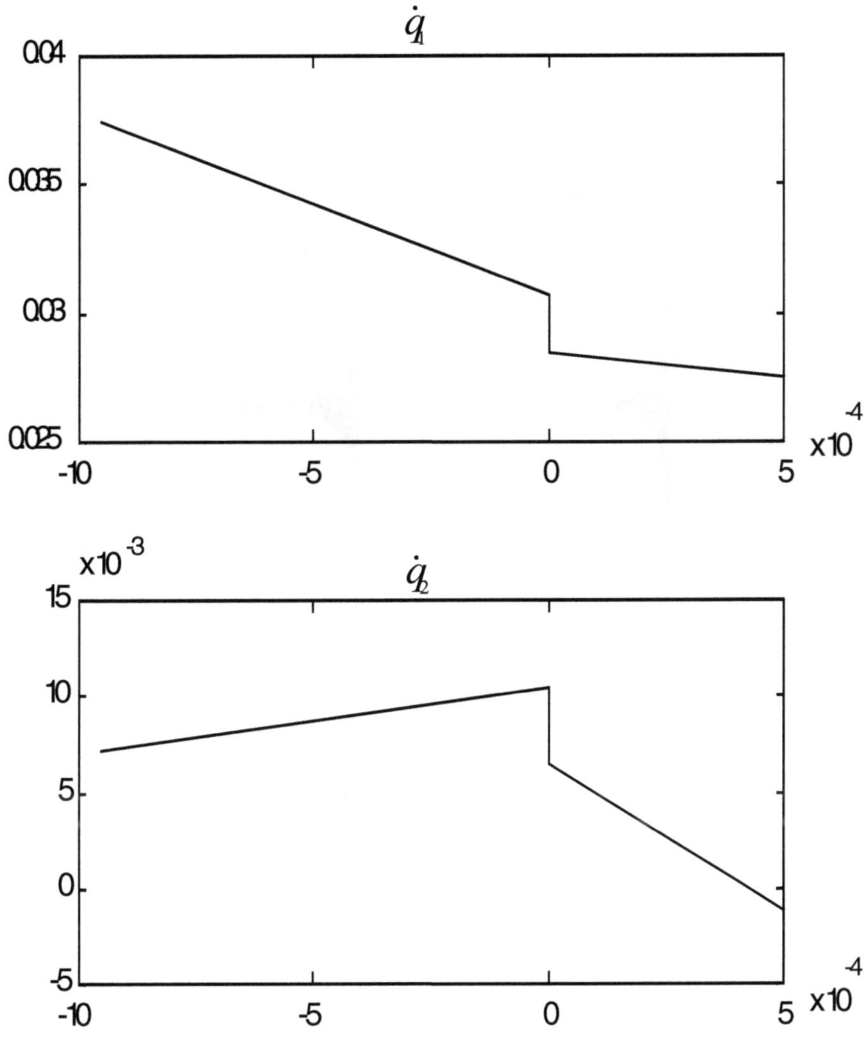

Figure 4.21. Impact (two-stage environment)
Time histories of the joint velocities show that sudden changes appear at the instant of impact. Impact instant is marked by 0.

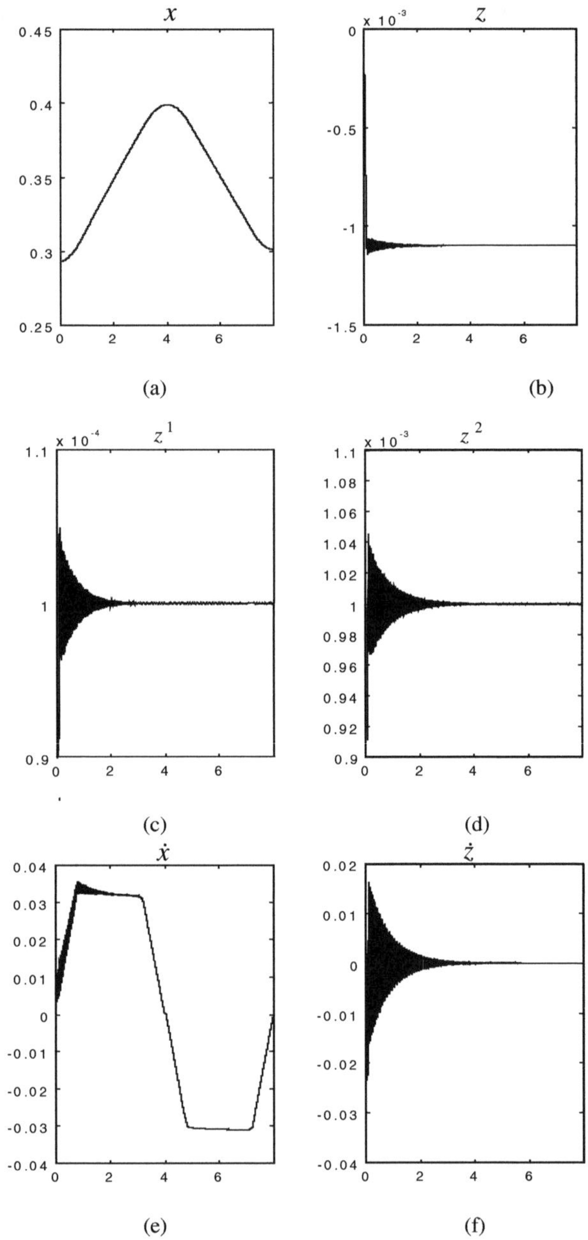

Figure 4.22. Simulation results: classical hybrid control and two-stage environment.
Motion of the robot tip during the contact phase:
(a), (b) – Time histories of the horizontal and vertical motion (x and z),
(c), (d) – external and internal components of deformation z (z^1 and z^2),
(e), (f) – velocities $\dot{x}$ and $\dot{z}$
External deformation is rather small compared to the internal one.
A long transient phase with oscillations may be observed.

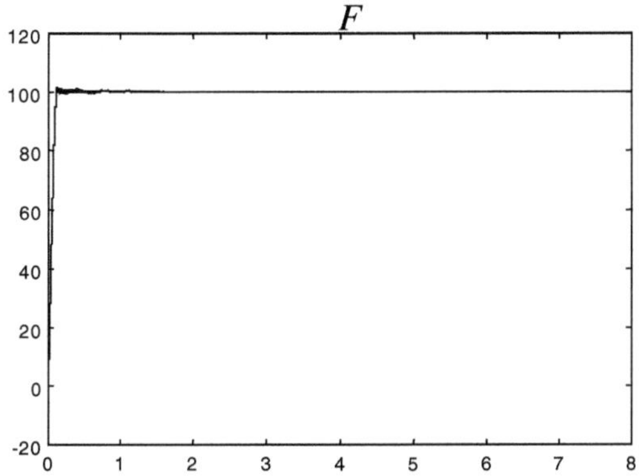

(a)

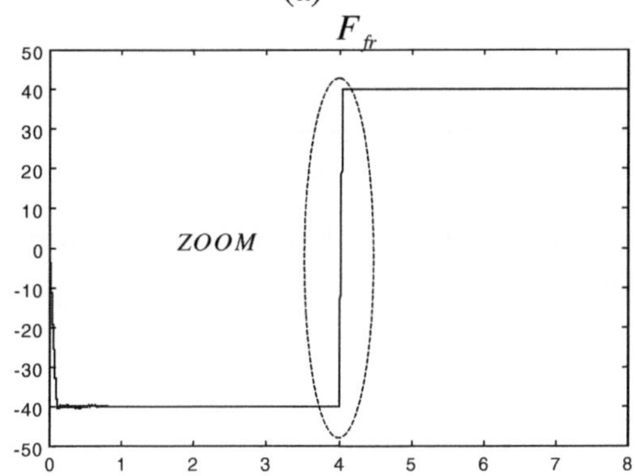

(b)

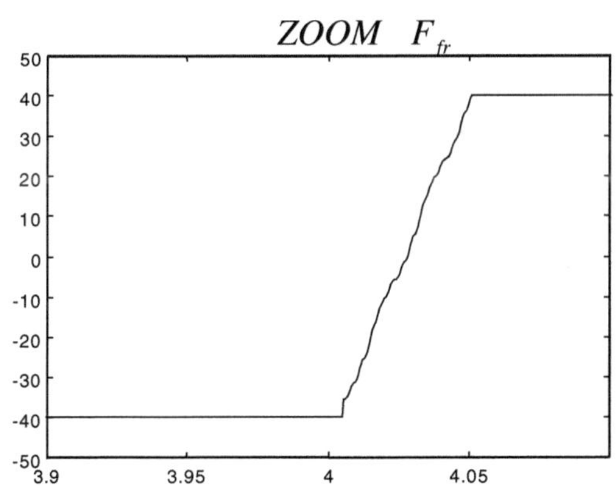

(c)

Figure 4.23. Simulation results: classical hybrid control and two-stage environment. Forces at the contact point: (a) – contact force F, and (b), (c) – friction force F_{fr}. Interval of jamming may be observed in (c).

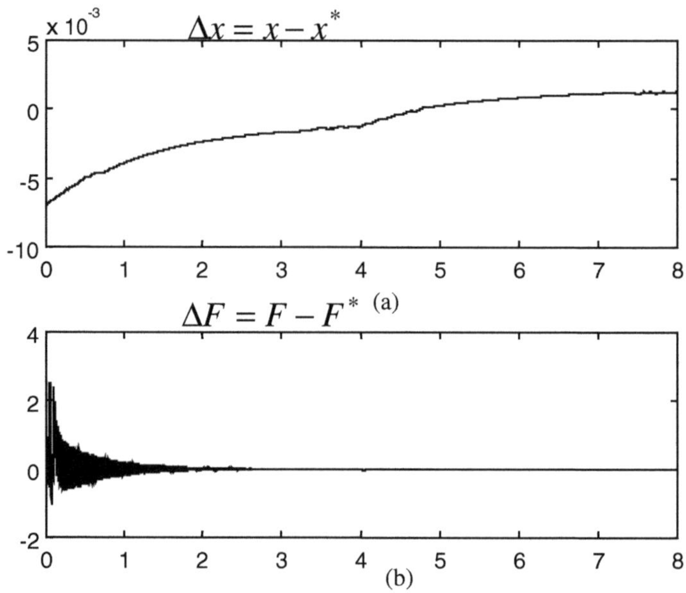

Figure 4.24. Simulation results: classical hybrid control and two-stage environment. Errors: (a) – position error Δx , and (b) – contact force error ΔF . Transient phase with oscillations may be observed.

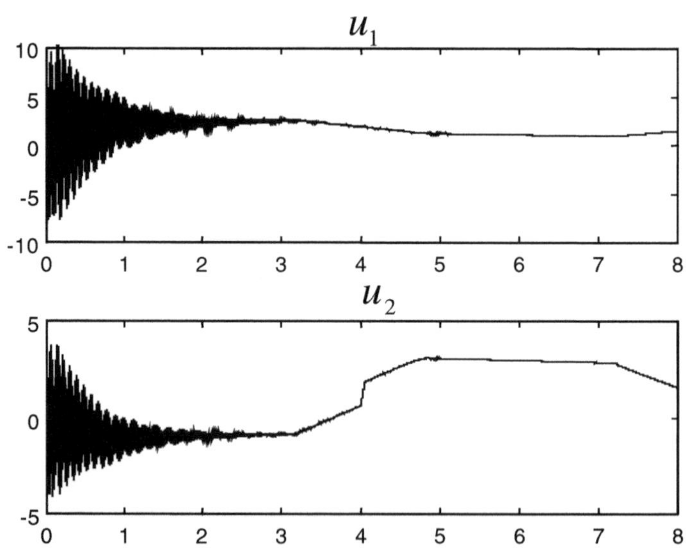

Figure 4.25. Simulation results: classical hybrid control and two-stage environment. Control voltages of DC motors
Long transient phase with oscillations may be observed.

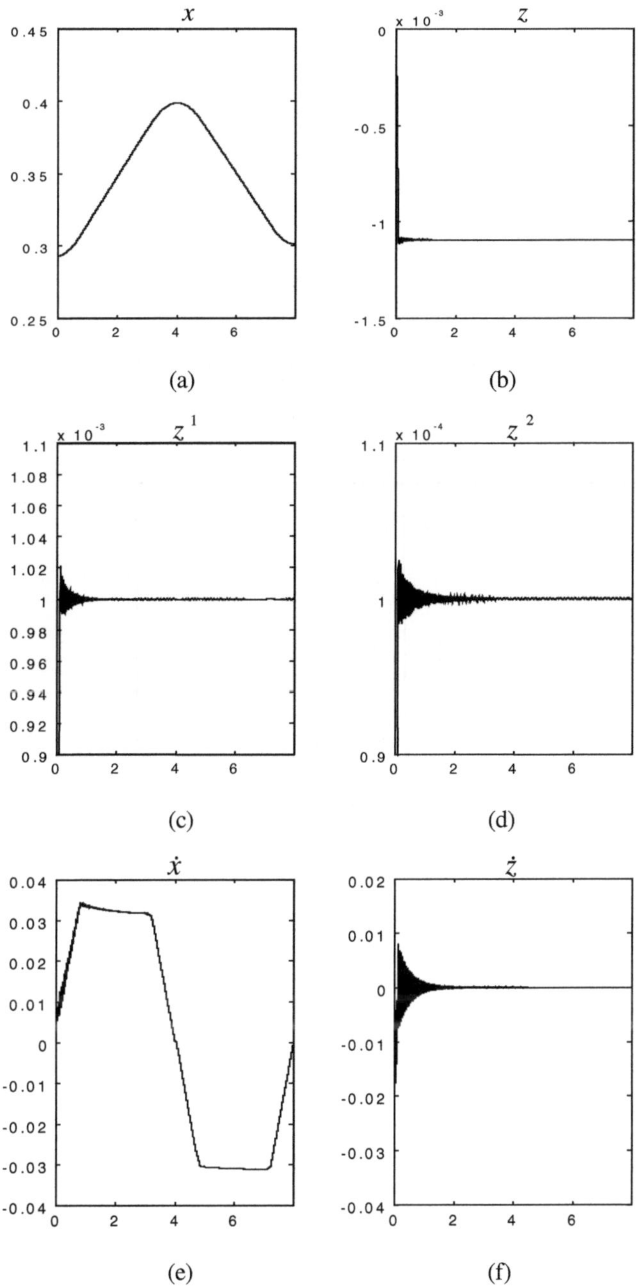

Figure 4.26. Simulation results: two-stage environment with smaller external stiffness. Motion of the robot tip during the contact phase:
(a), (b) – Time histories of the horizontal and vertical motions (x and z),
(c), (d) – external and internal components of deformation z (z^1 and z^2),
(e), (f) – velocities $\dot{x}$ and $\dot{z}$.

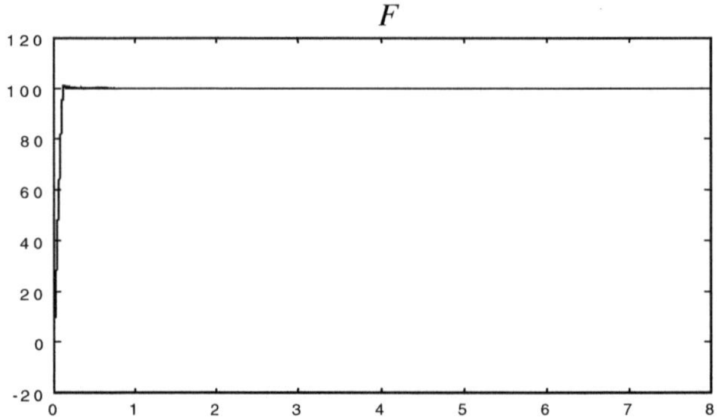

(a)

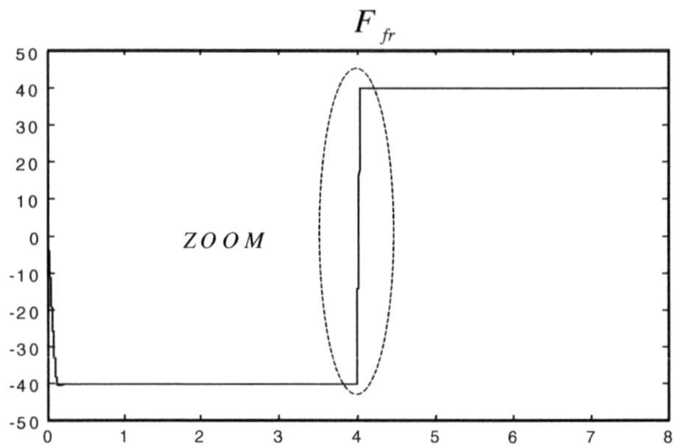

(b)

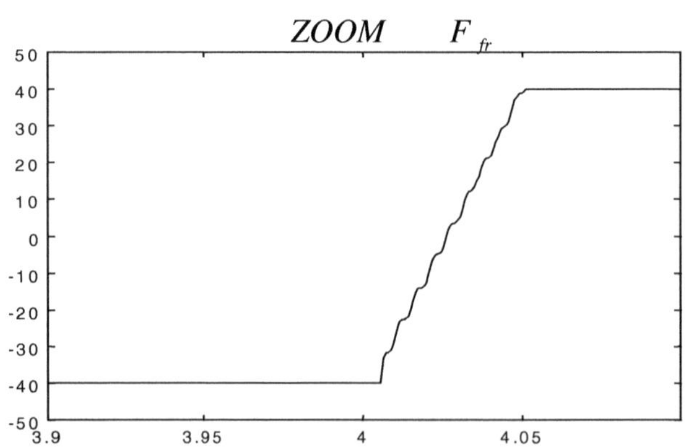

(c)

Figure 4.27. Simulation results: two-stage environment with smaller external stiffness. Forces at the contact point: (a) – contact force F, and (b), (c) – friction force F_{fr}. Interval of jamming may be observed in (c).

When the control law was changed and the dynamic position control applied, the results were improved. Oscillations in the transient phase were reduced and jamming was practically eliminated. Instead of presenting these results, we are going to examine a more interesting effect.

The presented simulation was performed under a reasonable assumption that the external stiffness is higher than the internal. It is a common case in the industry. However, in a specific robot task, a completely different situation may arise. Imagine a medical robot that performs massage. The human body is such that the external stiffness is considerably smaller than the internal. So, different combination of parameters should be used to simulate this situation. We adopted the following: $m^{(e)1} = 0.3\,kg$, $k^1 = 10^5\,N/m$, $b^1 = 1\,N/m/s$, $m^{(e)2} = 0.7\,kg$, $k^2 = 10^6\,N/m$, $b^2 = 1\,N/m/s$ (in fact, the values for k^1 and k^2 used in the previous simulation are now interchanged). The classical hybrid control is again applied. Figures 4.26–4.29 present the results of simulation. Results are rather similar to the previous ones. Only the extents of the external and the internal deformations changed (the values being almost exchanged).

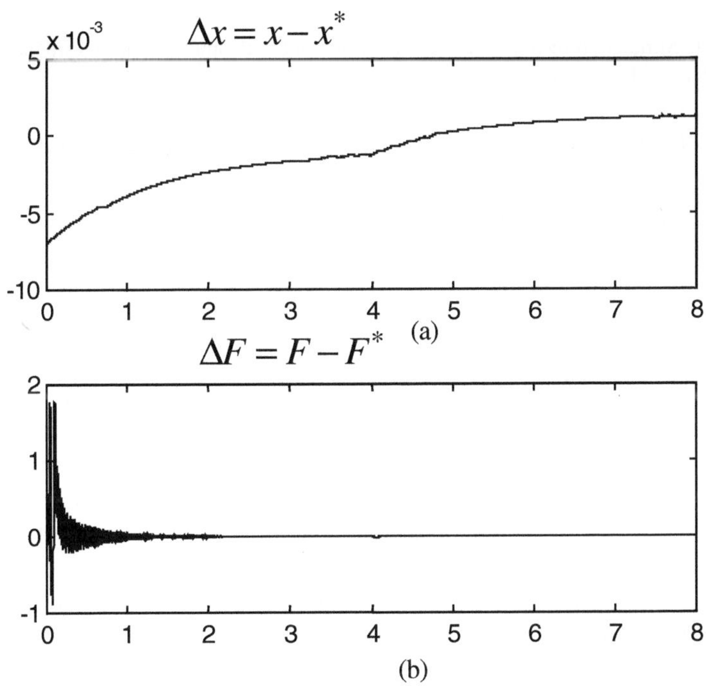

Figure 4.28. Simulation results: two-stage environment with smaller external stiffness. Errors: (a) – position error Δx, and (b) – contact force error ΔF.
Transient phase with oscillations may be observed.

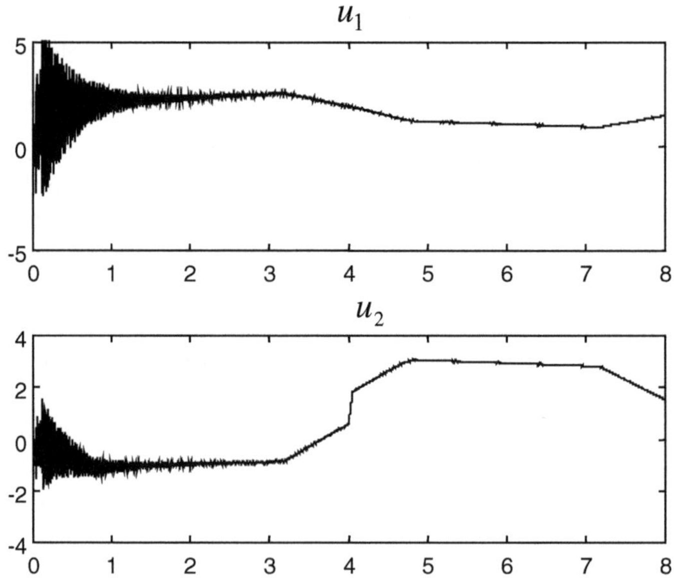

Figure 4.29. Simulation results: two-stage environment with smaller external stiffness. Control voltages of DC motors

2.3 Robot with Flexible Joints in a Contact Task

In the previous examples we have considered a robot with gearboxes assumed to be infinitely rigid. This means that no deformation existed between the motor and the joint shaft. Such an assumption may be considered justified in most practical situations. However, there are gearboxes which feature elastic properties and require more careful analysis. This is the case with Harmonic-Drive reducers. Elastic deformation (torsion) appears in the tube part of the device and should be taken into account if precise modeling is needed (torsion causes an error in angle transmission). With the robots that have motors dislocated from joints to the base (in order to achieve better static compensation), the torque transmission elements represent a new source of elastic deformation. Thus, there is a need to model and examine the elastodynamic effects coming from all the transmissions which exist between the motor and the corresponding robot joint. This problem, involving no constraints to the robot motion, was discussed in Section 2.3 in Ch 2. Here, we introduce constraints upon the motion. The contact makes the problem of transmission deformation even more important. This is due to possible interaction between the deformation of transmission and the deformation of the environment. This section provides two examples to demonstrate dynamic modeling of such system.

Example 1

Derivation of the model

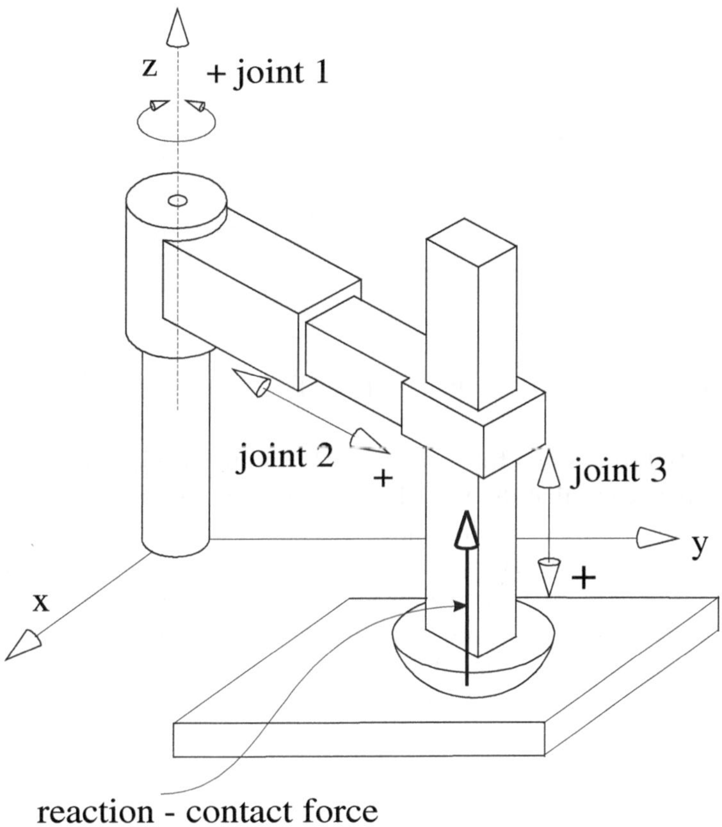

Figure 4.30. Cylindrical robot in a contact task. Subsystem of joint 3 may be dynamically decoupled from the rest of the system.

The first example considers a cylindrical robot engaged in a process operation which involves contact (Fig. 4.30). It can be a writing task, grinding, polishing, or other cutting operation. Our main objective is to examine the relationship between joint deformation and contact force. Hence, we are going to simplify the system preserving the relevant effects. The robot has three DOFs and accordingly three internal coordinates $q = [q_1\ q_2\ q_3]^T$. The functional coordinates are $s = [x\ y\ z]^T$. During approaching the surface (free motion), all coordinates are independent. When contact is established, the set s has to be partitioned into two subsets: $s^f = [x\ y]^T$ representing free motion along the surface, and $s^c = [z]$ being the constrained coordinate. For the robot mechanism without transmissions and actuators, the dynamic model may be written with no flexibility at the joints involved. The dynamic model would have the form (3.11). For the particular configuration shown in Fig. 4.30, the structure of the model (3.11) becomes:

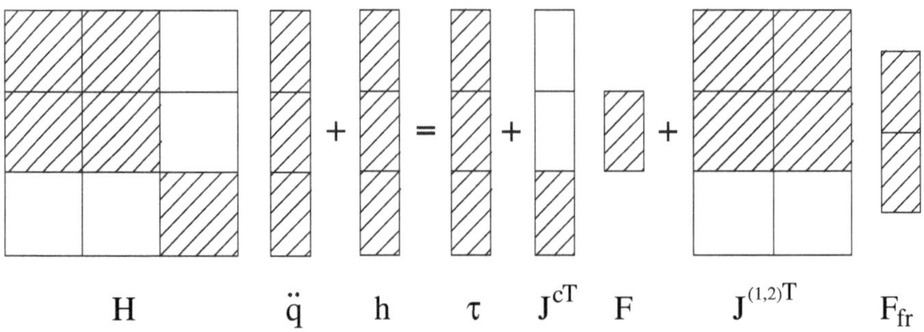

where empty boxes stand for zero elements and the shaded ones for nonzero elements. It is clear that the model can be divided into two submodels: one containing the first two scalar equations (from the matrix model) and the other containing the third equation. These models are still coupled. The third equation influences the first two through the friction force (F_{fr} depends on F). However, the backward influence does not exist. The first two equations do not affect the third one. Thus, the model describing the dynamics of q_3 is decoupled. We now concentrate on this subsystem only. It has the form

$$H_{33}\,\ddot{q}_3 + h_3 = \tau_3 + \frac{\partial z}{\partial q_3} F \,. \tag{4.17}$$

Coefficient H_{33} represents the mass of link 3 (let it be marked by m), and h_3 represents the gravity force mg . The Jacobian is $\partial z / \partial q_3 = -1$. Since we consider q_3 only (that is, we do not discuss q_1 and q_2), index "3" in (4.17) may be omitted and the equation rewritten as:

$$m\ddot{q} - mg = \tau - F \,. \tag{4.18}$$

The gravity term can be omitted without losing any relevant effect. Let us justify this. First, with the manipulators of considered structure, gravity is usually compensated. Second, the gravity term appears only if the link moves in the vertical direction. If the motion is horizontal, gravity does not influence it, and all the effects in which we are interested will still be present. Hence, in the consideration that follows, gravity will be ignored.

Let us now define the drive and the transmission for the joint considered. The joint is driven by a DC motor with the following parameters: $I = 7.9 \cdot 10^{-4} kgm^2$, $R = 0.175\Omega$, $C_M = 0.0575 Nm/A$, $C_E = 0.0575 V/rad/s$. The transmission is shown in Fig. 4.31. Its main part is a Harmonic-Drive (HD) with the gear ratio $N = 50$ and the moments of inertia I', I'' (according to Section 2.3 in Ch. 2). In the numerical simulation, these inertial moments will be neglected (they are small compared to the rotor inertia and the inertia of the link). The tube part of HD is considered deformable (torsion), with the stiffness $c = 14000\, Nm/rad$ and damping $d = 0.0196\, Nm/rad/s$. This is the only source of deformation in the transmission. The wheel following after HD turns rotation into translation. Its radius is $r = 0.15\, m$ and its inertia is immediately incorporated into the link mass.

Mass of the link is $m = 50\, kg$.

Now we talk about the dynamics of the transmission; first about the HD. According to the scheme from Fig. 4.31, θ^1 is the input angle and M^1 the input torque for HD. They coincide with the motor shaft angle and its output torque: $\theta^1 = \theta$ and $M^1 = M$. Let θ^2 and M^2 be the output variables of HD. The angle θ^2 is proportional to the link translation, $\theta^2 = q/r$, and the torque is proportional to the link driving force, $M^2 = \tau r$. According to the theory given in Section 2.3 in Ch. 2 the dynamics of transmission can be described by:

$$(I' + I + I''/N^2)\ddot{\theta} = M^1 - M^2/N, \tag{4.19}$$

and the motor dynamics by:

$$M^1 = M = \frac{C_M}{R}u - \frac{C_M C_E}{R}\dot{\theta}. \tag{4.20}$$

The HD output torque deforms the tube part, and hence, this can be expressed as:

$$M^2 = \tau r = c\left(\frac{\theta}{N} - \frac{q}{r}\right) + d\left(\frac{\dot{\theta}}{N} - \frac{\dot{q}}{r}\right). \tag{4.21}$$

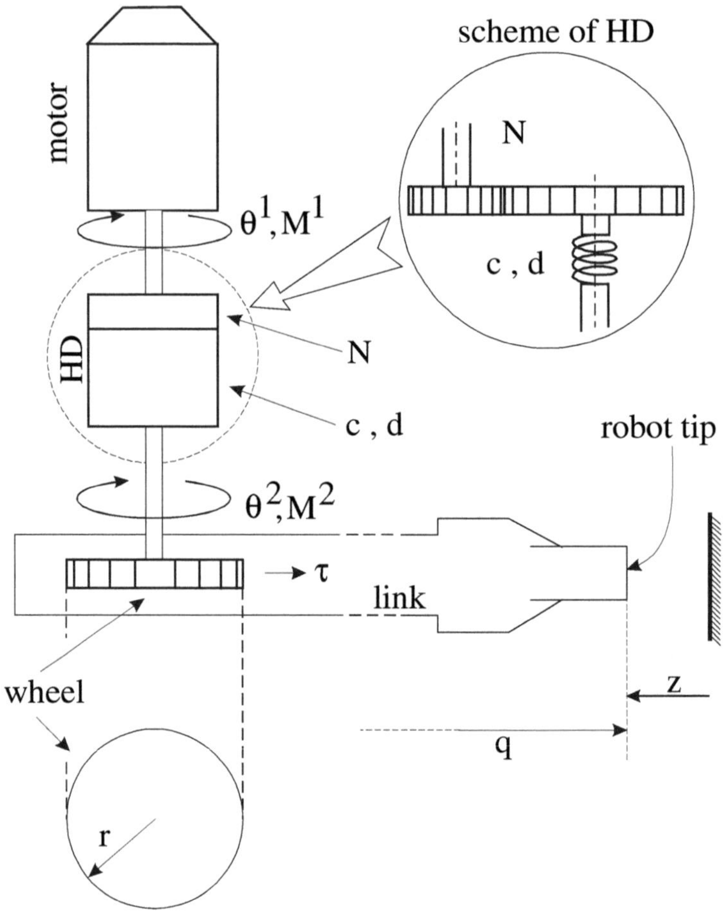

Figure 4.31. One-dimensional robot: coordinates, drive, and transmission.

Equations (4.19)–(4.21) can be combined do give

$$\left(I + I' + \frac{I''}{N^2}\right)\ddot{\theta} = \frac{C_M}{R}u - \frac{C_M C_E}{R}\dot{\theta} - \frac{c}{N}\left(\frac{\theta}{N} - \frac{q}{r}\right) - \frac{d}{N}\left(\frac{\dot{\theta}}{N} - \frac{\dot{q}}{r}\right) \tag{4.22}$$

If the link driving force τ is substituted from (4.21) into (4.18), the dynamic model of the link acquires the form:

$$m\ddot{q} = \frac{c}{r}\left(\frac{\theta}{N} - \frac{q}{r}\right) + \frac{d}{r}\left(\frac{\dot{\theta}}{N} - \frac{\dot{q}}{r}\right) - F\,, \tag{4.23}$$

where the gravity term was omitted. Equations (4.22) and (4.23) describe the robot dynamics. It is a one-dimensional robot that has two DOFs due to elastic deformation in the transmission. Position coordinates are q for the link and θ for the motor, and they are kinematically independent. If the robot were free (no contact), the force F would not appear in the model (4.23). This would apply to the approaching phase of the contact task. The contact is established when the functional coordinate z becomes zero (Figs. 4.31, 4.32).

After the impact, the contact motion starts. The dynamics of the environment is now needed to complete the overall model. As already stated, the environment (surface) may be modeled by means of lumped masses. Here, we concentrate again on the systems with one-stage deformation and two-stage deformation. The first case is shown in Fig. 4.32. One may note that the deformation z depends directly on q. If q_e is the value at the instant of collision (z=0), then:

$$z = q_e - q \,. \tag{4.24}$$

Note that z is negative.

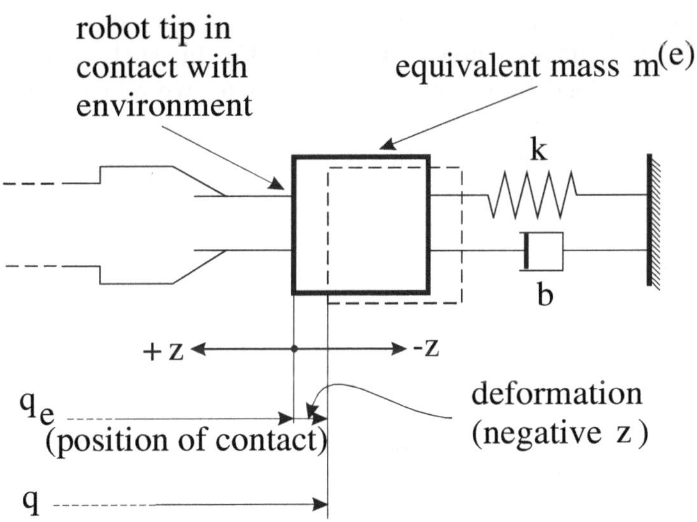

Figure 4.32. One-stage deformation of the surface.

The environment dynamics can now be described by Newton's law applied to the equivalent mass $m^{(e)}$, that is the model (4.15). If z is replaced by (4.24), the model (4.15) becomes:

$$F = m^{(e)}\ddot{q} + kq + b\dot{q} - kq_e \,. \tag{4.25}$$

Equations (4.22), (4.23), and (4.25) define the overall robot-plus-environment dynamics. These three equations contain three unknowns: two accelerations ($\ddot{q}$ and $\ddot{\theta}$) and the contact force F. Thus, simulation is possible.

We now return to the instant of collision and consider the impact. As we did in the previous sections, we consider the impact as nonelastic and infinitely short. The position coordinates q and θ do not change, but the velocity $\dot{q}$ has a finite change at that instant (the motor speed $\dot{\theta}$ does not change, since transmission elasticity prevents the impact shock). Let $\dot{q}'$ be the velocity before the impact and $\dot{q}''$ the new value after impact. In this simple case, the theory explained for impact reduces to the law of momentum conservation:

$$m\dot{q}' = (m + m^{(e)})\dot{q}'' . \tag{4.26}$$

In this way, the initial state for the contact dynamics is found.

For the purpose of simulation we adopt the following characteristics of the environment: $m^{(e)} = 10\,kg$, $k = 10^5\,N/m$, and $b = 0.5\,N/m/s$.

Now, we consider a more complex model of the environment deformation, the two stages shown in Fig. 4.33. In such case the elastodynamics is described by the model (4.16). Here, we adopt that the external equivalent mass is small enough to be neglected $(m^{(e)1} = 0)$, so that only one mass $(m^{(e)2} = m^{(e)})$ remains. Model (4.16) may now be simplified to give:

$$\begin{aligned}
&F = -k^1 z^1 - b^1 \dot{z}^1 \\
&k^1 z^1 + b^1 \dot{z}^1 = m^{(e)} \ddot{z}^2 + k^2 z^2 + b^2 \dot{z}^2 \\
&z = z^1 + z^2
\end{aligned} \tag{4.27}$$

Equations (4.22) and (4.23) define the robot dynamics, while (4.27) describes the dynamics of the environment. Expression (4.24) still holds and relates the coordinate q with the total deformation z. In this way we come to the overall dynamic model. The system has 3 DOFs and three coordinates should be selected to define the position (e.g., q, θ, z^1 or q, θ, z^2). The model can be solved for the system accelerations and the contact force, thus allowing simulation.

Since the external equivalent mass does not exist, the impact does not exist either. After the collision, the contact motion starts immediately. This does not mean that there is no transient phase before the "regular" contact behavior is established.

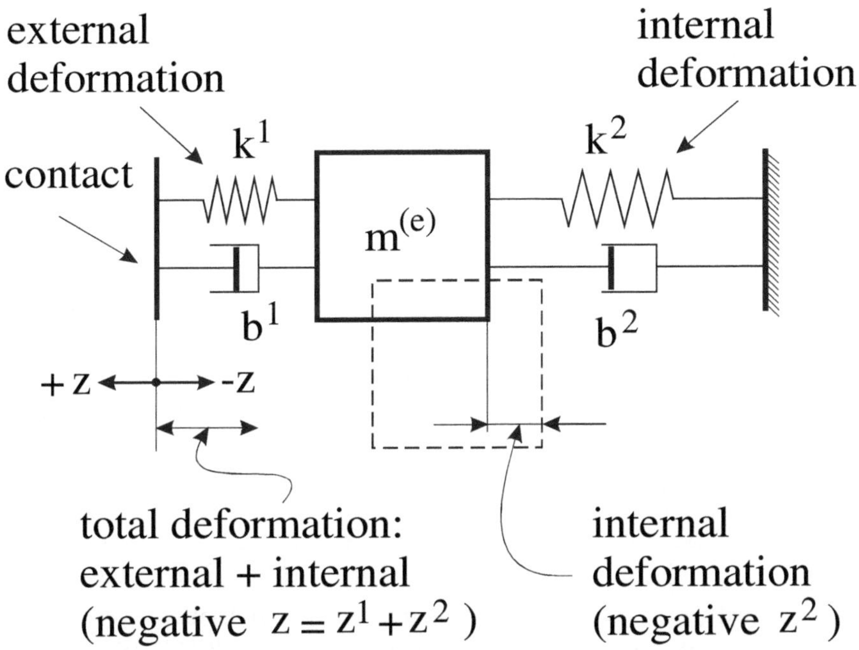

Figure 4.33. Two-stage deformation of the surface.

For the purpose of simulation we adopt the following characteristics of the environment: $m^{(e)} = 10\,kg$, $k^1 = k^2 = 10^5\,N/m$, $b^1 = 1\,N/m/s$, and $b^2 = 0.5\,N/m/s$.

Simulation results

The task consists of the approaching phase and contact motion. The robot starts from the position $q(0) = 0$, and the distance to the surface is $q_e = 1m$. Approaching should be accomplished in T_1=1.4 s with a trapezoidal velocity profile. Motion is planned so as to avoid the impact (zero velocity at the instant of contact). In this phase the robot is controlled by a PD regulator. The control does not take care of transmission flexibility, as the precise tracking during approaching is not considered important in this task. The tracking error will cause collision with the surface (that is, nonzero-velocity contact) and impact will occur. The tracking error $\Delta q = q - q*$ is shown in Fig. 4.34.

In the contact phase, the robot has to produce a pressure force $F = 100\,N$ upon the surface. The force should increase linearly from 0 to 100 N in 0.1 s and then remain constant. For the control of contact behavior, two versions of dynamic control of robot were checked: the first control law was based on information regarding position and the other on information regarding force [3–5] (for control issues, see Appendix 2).

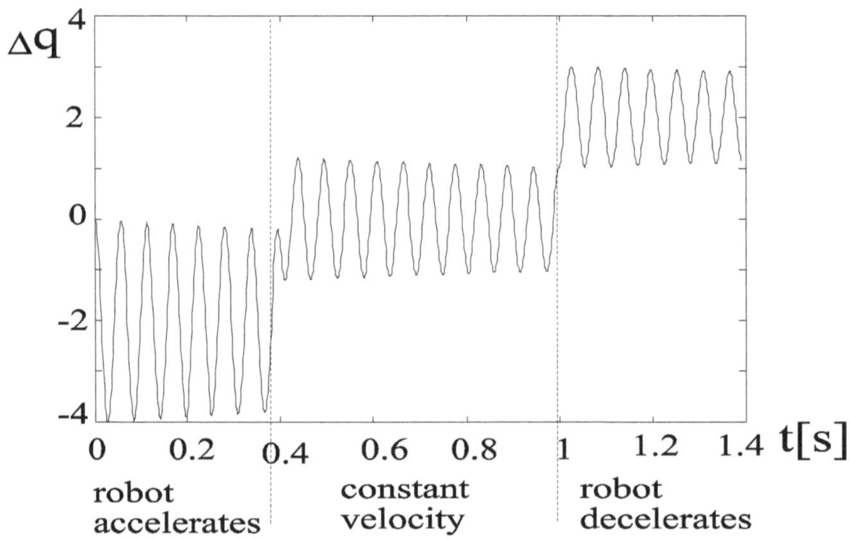

Figure 4.34. Tracking error $\Delta q = q - q^*$ in the approaching phase. Effects due to the elastic deformation in transmission may be observed.

Figs. 4.35–4.37 present the simulation results. In Fig. 4.35 the errors in force and position are shown for the case of dynamic position control and the one-stage environment model. The same control concept was applied to the robot in contact with two-stage environment and the results are shown in Fig. 4.36. One may find that with the one-stage environment (Fig. 4.35) the transient phase with large variations in force and position is strongly expressed. With the two-stage model, the transient phase could not be observed. This is due to the fact that there was no mass exposed to impact $(m^{(e)1} = 0)$.

The dynamic force control was applied to the one-stage environment only and the simulation results are presented in Fig. 4.37. The first two diagrams, obtained for the parameters defined above (that is $m^{(e)} = 10\,kg$, $k = 10^5\,N/m$, and $b = 0.5\,N/m/s$), demonstrate the acceptable behavior of the contact system. However, variations of the environment parameters showed that dynamic force control may sometimes lead to inappropriate results in position tracking. This is evident from the latter two diagrams of Fig. 4.37 (the parameters used: $m^{(e)} = 5\,kg$, $k = 10^3\,N/m$, and $b = 0$).

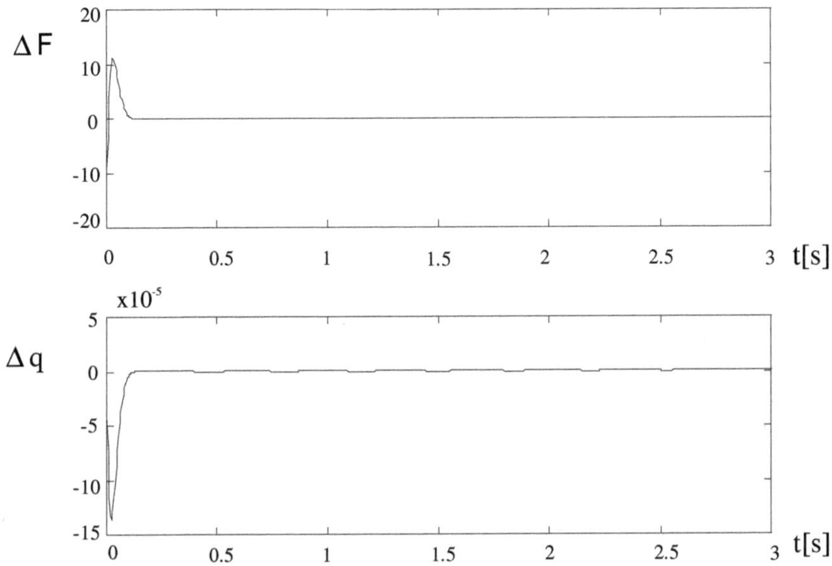

Figure 4.35. Simulation – dynamic position control applied to one-stage environment: Errors in force (ΔF) and in position (Δq).

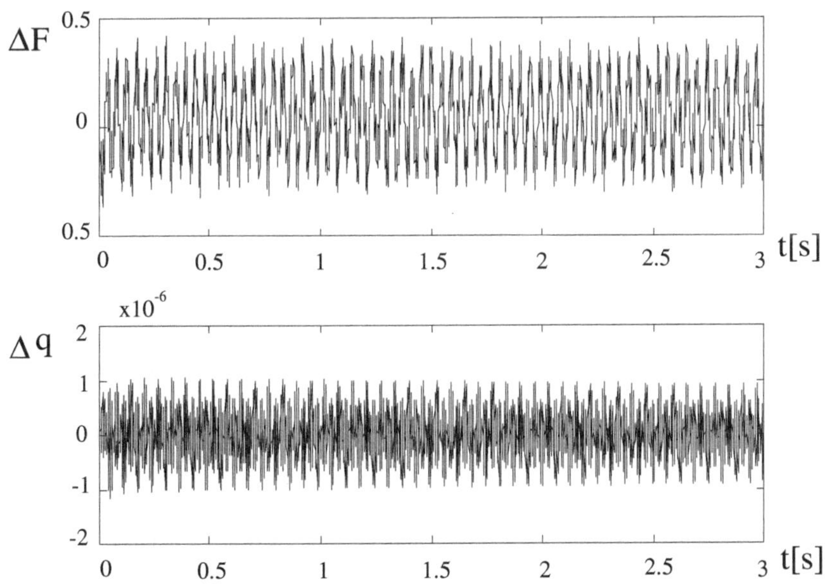

Figure 4.36. Simulation – dynamic position control applied to two-stage environment: Errors in force (ΔF) and in position (Δq).

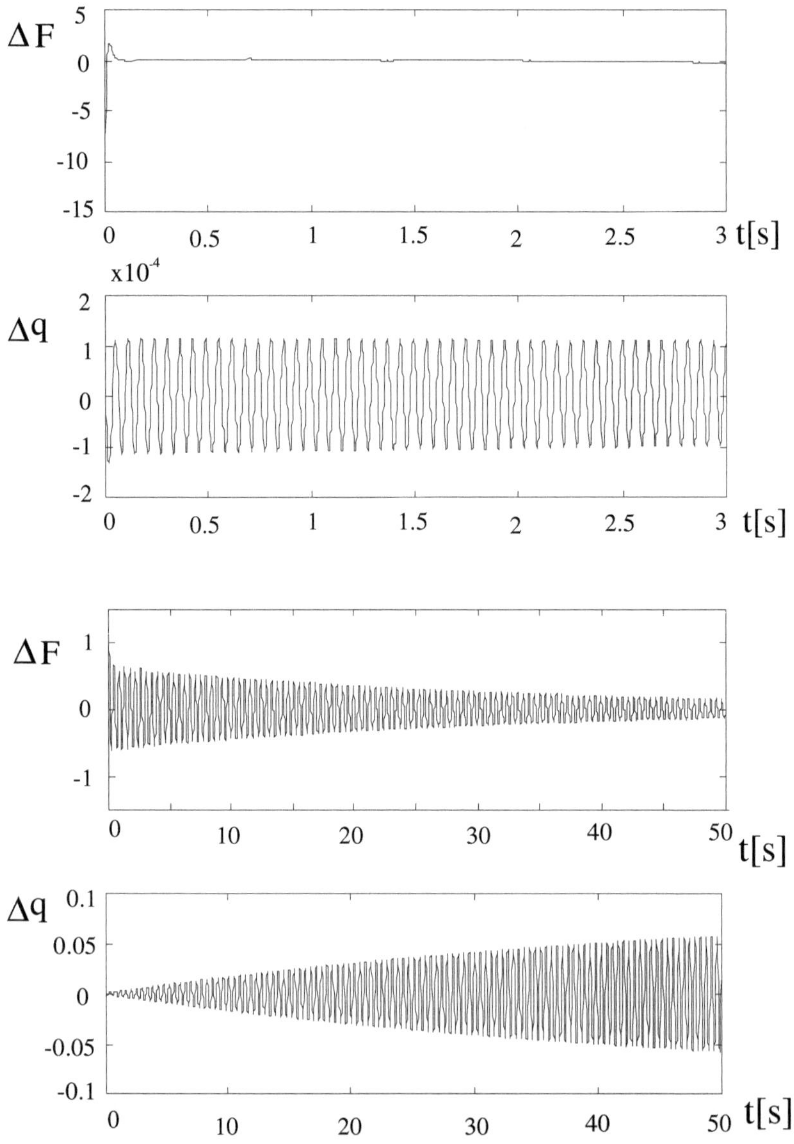

Figure 4.37. Simulation – dynamic force control applied to one-stage environment: Force and position errors $(\Delta F, \Delta q)$ for two different sets of the environment parameters. The first two diagrams demonstrate acceptable behavior, while the latter two demonstrate inappropriate tracking of position.

Example 2

Let us consider another example, a two-link planar robot with revolute elastic joints engaged in a writing task. Description of the robot and the task are shown in Fig. 3.6. The only difference from the example presented in Fig. 3.6. is that robot joints are not absolutely rigid this time. The sources of elasticity are HD reducers in robot joints. Approaching phase (A→B), and impact were considered in previous text and this time we will consider only contact phase of the motion (B→C).

During contact phase of the motion, dynamics of robot interacting with dynamic environment is described by a system of nonlinear differential equations:

$$H(q)\ddot{q} + h(q,\dot{q}) = \tau + J^T F \tag{4.28}$$

where q is a 2-dimensional vector of generalized coordinates (joint coordinates), H is a 2×2 positive definite matrix of mechanism inertia moments, h is a 2-dimensional nonlinear vector function of the centrifugal, Coriolis' and gravitational moments, τ is a 2-dimensional vector of the driving torques, J is a 2×2 matrix of Jacobi (connecting the velocities of the robot end-effector and the velocities of robot generalized coordinates), and F is a 2-dimensional vector of contact forces.

Each joint ($i = 1,2$) is driven by a DC motor with corresponding transmission ratio. We consider Harmonic-Drive reducers as transmission units. An HD reducer can be modelled as a rigid-gear reducer plus a torsional spring (Fig. 4.38), where θ_i is the motor shaft angle, q_{1i} is the joint shaft angle, c_i is torsional elasticity, and d_i is torsional damping. The number of degrees of freedom (DOF) is doubled, since each mechanism joint obtains supplementary (passive) DOF due to the elasticity of the transmission unit. The equations describing the dynamics of motor and reducer for the joint shown in Fig. 4.38 can be written in the following form:

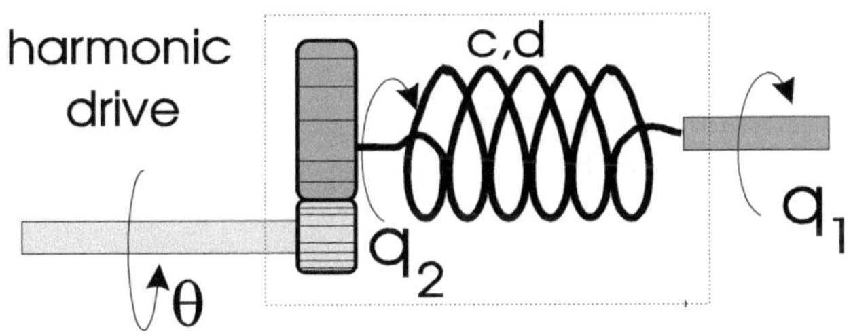

Figure 4.38. Model of Harmonic-Drive gearbox.

$$u_i = R_i i_i + C_{Ei}\dot{\theta}$$

$$C_{M_i} i_i = J_{ri}\ddot{\theta}_i + B_i\dot{\theta}_i + \frac{\tau_i}{N_i}, \tag{4.29}$$

$$\tau_i = c_i\left(\frac{\theta_i}{N_i} - q_{1i}\right) + d_i\left(\frac{\dot{\theta}_i}{N_i} - \dot{q}_{1i}\right) = c_i(q_{2i} - q_{1i}) + d_i(\dot{q}_{2i} - \dot{q}_{1i})$$

where u_i is the input control voltage, R_i is resistance, i_i is motor current, C_{M_i} and C_{E_i} are constants of torque and e.m.f., respectively, J_{ri} is rotor and reducer moment of inertia, B_i is viscous friction coefficient, N_i is reduction ratio of reducer rigid part, and τ_i is the torque on joint shaft. Coefficients c_i and d_i describe stiffness and damping of the HD reducer, respectively.

This set of equations, applied for both joints, acquires matrix form:

$$\begin{gathered} u = Ri + C_E\dot{\theta}\ ,\ C_M i = J_r\ddot{\theta} + B\dot{\theta} + N^{-1}\tau, \\ \tau = C\left(N^{-1}\theta - q\right) + D\left(N^{-1}\dot{\theta} - \dot{q}\right) = C(q_2 - q_1) + D(\dot{q}_2 - \dot{q}_1) \end{gathered} \tag{4.30}$$

where $u = diag\{u_i\}$, $R = diag\{R_i\}$, $i = diag\{i_i\}$, $J_r = diag\{J_{ri}\}$, $\theta = diag\{\theta_i\}$, $q_1 = diag\ \{q_{1i}\}$, $q_2 = diag\ \{q_{2i}\}$, $C_E = diag\ \{C_{E_i}\}$, $C_M = diag\ \{C_{M_i}\}$, $C = diag\ \{c_i\}$, $D = diag\{d_i\}$, and $N = diag\ \{N_i\}$.

In order to complete the description of dynamics, mathematical model of environment is needed. We shall consider dynamic environment presented in Fig 4.32 and described by equation 4.25.

We will examine three types of control laws: classical hybrid position/force control (see Ref. [2] and Appendix 1), dynamic position (see Ref. [3, 4] and Appendix 2), and dynamic force control laws (see Ref. [3–5] and Appendix 2). Classical hybrid control has already been used in examples in previous text and dynamic position and force control will be presented in the text to follow.

Dynamic Position Control

The idea of dynamic position control is based on the fact that the contact force and the manipulator tip position are connected by equation (4.25), so that the contact force can be controlled by controlling the mechanism tip position. The desired position i.e. trajectory of the mechanism, corresponding to the desired contact force, is calculated from the equation of the environment (4.25):

$$m\ddot{z}^o = F - b\dot{z}^o - k(z^o - z_o) \tag{4.31}$$

In x-direction, we should specify nominal motion x^o. Complete nominal motion vector can be written in the following way:

$$X = \begin{bmatrix} x^o \\ z^o \end{bmatrix} \tag{4.32}$$

From inverse kinematics equation we can calculate joint motions:

$$\ddot{q}^o = J^{-1}(\ddot{X} - A)$$

where J is Jacobian matrix.

The control should ensure that $q(t) \to q^o(t)$ and $\dot{q}(t) \to \dot{q}^o(t)$. Position control is based on dynamic position control and can be written in the following way:

$$\begin{aligned} u = RC_M^{-1}\{ & J_r(\ddot{q}_2^o + P_2(\Delta q_2, \Delta \dot{q}_2)) + C_M C_E R^{-1} \dot{q}_2 + \\ & + N^{-1}[C(q_2 - q_1) + D(\dot{q}_2 - \dot{q}_1)]\} \end{aligned} \tag{4.33}$$

where q_2^o can be calculated from the following equation:

$$\begin{aligned} \dot{q}_2^o = \dot{q}_1^o + D^{-1}\{ & H(q_1)[\ddot{q}_1^o + P_1(\Delta q_1, \Delta \dot{q}_1)] + h(q_1, \dot{q}_1) - \\ & - C(q_2 - q_1) - J^T \begin{bmatrix} -\mu sign(\dot{x}^o)(m\ddot{z} + b\dot{z} + k(z - z_o)) \\ m\ddot{z} + b\dot{z} + k(z - z_o) \end{bmatrix} \} \end{aligned} \tag{4.34}$$

Since the second derivation q_2 is also needed, it is obtained by numerical differentiating equation (4.34).

Dynamic Force Control

When dynamic force control is applied, besides the desired contact force, the character of the transient response during the realization of this contact force is also set. This transient response with respect to force is set by the following equation:

$$F(t) = F^o(t) + \int Q(\mu(t))dt \tag{4.35}$$

where F^o is the set nominal force; $\mu = F^o - F$ and Q is corresponding vector function.

On the basis of the environment model (4.25) it follows:

$$m\ddot{z}^o = F - b\dot{z}^o - k(z^o - z_o) \tag{4.36}$$

and from inverse kinematics equation we can calculate joint motions:

$$\ddot{q}^o = J^{-1}(\ddot{X} - A) \tag{4.37}$$

Force control relies on the force control law

$$\begin{aligned} u = RC_M^{-1}\{J_r(\ddot{q}_2^o + P_2(\Delta q_2, \Delta\dot{q}_2)) + C_M C_E R^{-1}\dot{q}_2 + \\ + N^{-1}[C(q_2 - q_1) + D(\dot{q}_2 - \dot{q}_1)]\} \end{aligned} \tag{4.38}$$

By considering a family of transient responses with respect to force in the form: $\dot{\mu} = Q(\mu)$ and by choosing continuous function Q(Q(0)=0) such that asymptotic stability as a whole is ensured for the trivial solution of $\mu(t)\equiv 0$, the stabilizing $F^o(t)$ could be solved. Similar to position control, force control calculates the trajectory q_2 , on the basis of the error between the desired and realized force, and after that performs stabilization of motor angle q_2 around q_2^o. Calculation of q_2^o is carried out by substituting equation (4.35) into (4.36) and putting the newly-obtained expression for acceleration into equation (4.29):

$$\begin{aligned} \dot{q}_2^o = \dot{q}_1 + D^{-1}\{H(q_1)\ddot{q}_1 + h(q_1, \dot{q}_1) - C(q_2 - q_1) - \\ -J^T(F^o(t) + \int Q(\mu(t))dt)\} \end{aligned} \tag{4.39}$$

Since the second derivation $\ddot{q}_2^o$ is also needed, it is obtained by numerical differentiating equation (4.39).

Simulation results

In the following text we shall present simulation results of two-link robot with elastic joints performing a contact task. Description of the robot and the task are shown in Fig. 3.6. The only difference from the example presented in Fig. 3.6. is that robot joints are not absolutely rigid. The sources of elasticity are HD reducers in robot joints. Let us adopt that in both robot joints, there are equal HD reducers with parameters as in Example 1. We shall pay attention to the contact phase of the motion only. In the contact phase of the task, the robot has to produce a pressure force $F = 100\,N$ upon the surface. The force should increase linearly from 0 to 100 N in 0.1 s and then stay constant. In x direction robot starts from the position x=0.3m and stops at position x=0.4m. Complete task should be executed in 1s with a trapezoidal velocity profile.

Figures 4.39–4.42 present the results of simulation. In Fig. 4.39 are presented simulation results when classical hybrid control is applied. For the purpose of simulation we adopt the following characteristics of the environment: $m = 1kg$, $k = 10^5 N/m$, $b = 1 N/m/s$. From this figure we notice that large position error in x direction exists as a consequence of friction and applied classical hybrid control. In Fig. 4.40 are presented simulation results when dynamic position control is applied. We have adopted the same environment parameters. In Fig. 4.41 are presented simulation results when dynamic force control is applied. Environment parameters are: $m = 1kg$, $k = 10^5 N/m$, $b = 10 N/m/s$. As we have already mentioned, if dynamic force control is used there are cases (depending on environment parameters) when force is stabilized while position has inappropriate behavior, as presented in Fig. 4.42. For this simulation adopted environment parameters are $m = 1kg$, $k = 10^5 N/m$, $b = 1 N/m/s$.

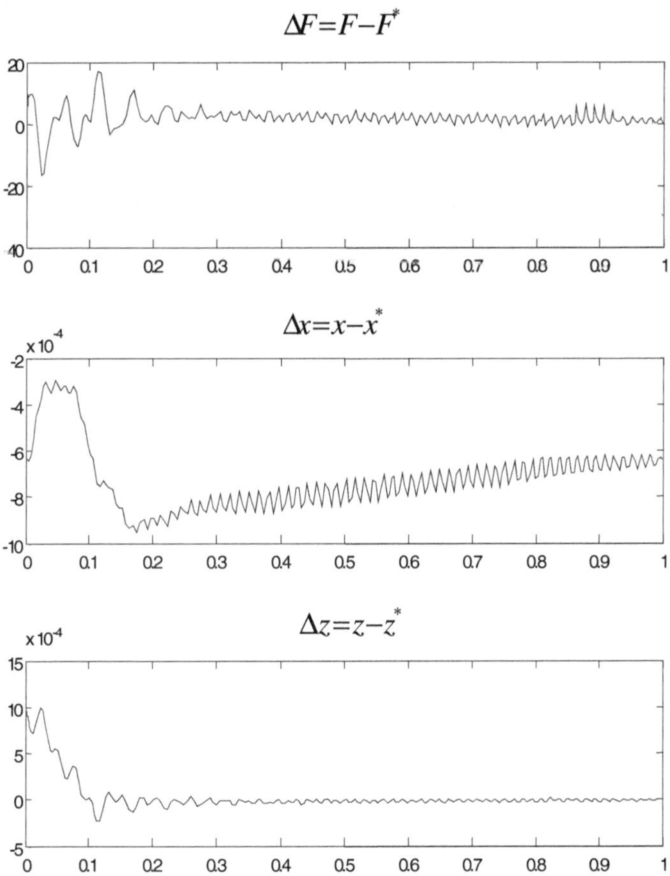

Figure 4.39. Simulation – classical hybrid position/force control applied to one-stage environment: Contact Force in z direction and position errors in x and z direction. Large position error in x direction as a result of classical hybrid control. Environment parameters are: $m = 1kg$, $k = 10^5 N/m$, $b = 1 N/m/s$.

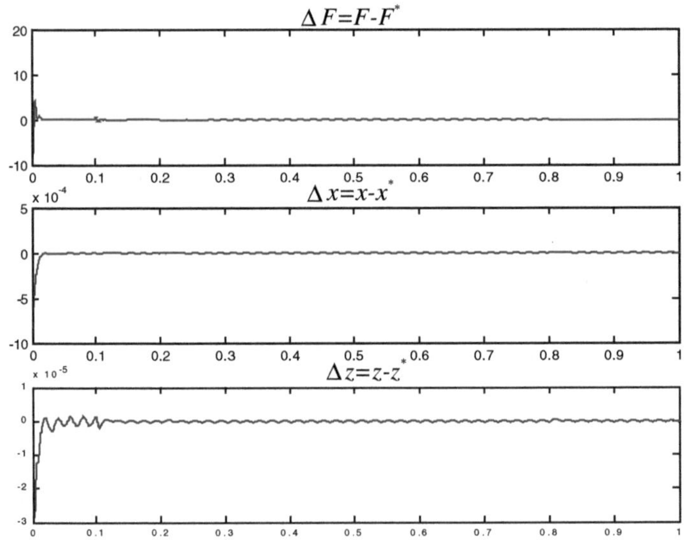

Figure 4.40. Simulation – dynamic position control applied to one-stage environment: Contact force error in z direction and position errors in x and z direction . Environment parameters are: $m = 1kg$, $k = 10^5 N/m$, $b = 1 N/m/s$.

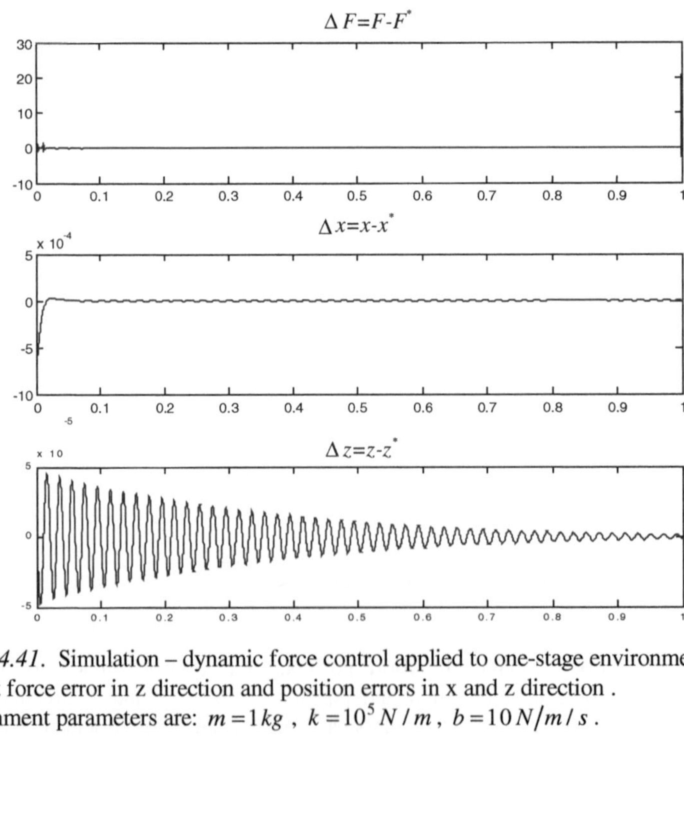

Figure 4.41. Simulation – dynamic force control applied to one-stage environment. Contact force error in z direction and position errors in x and z direction . Environment parameters are: $m = 1kg$, $k = 10^5 N/m$, $b = 10 N/m/s$.

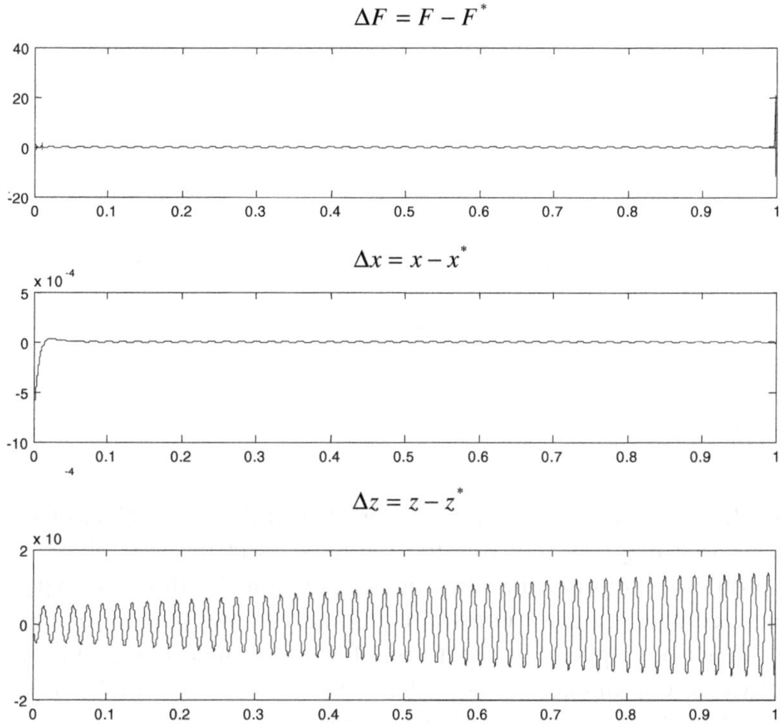

Figure 4.42. Simulation – dynamic force control applied to one-stage environment. Inappropriate behaviour. Environment parameters are: $m = 1\,kg$, $k = 10^5\,N/m$, $b = 1\,N/m/s$.

2.4 Flexible-Link Robots in Contact Tasks

In this section, flexibility of robot links will be introduced and the interaction with contact dynamics modeled. This does not have the same interest for conventional manipulation robots working in industry or services. In these robots, links are robust enough to prevent any considerable deformation. However, in space robotics (e.g., the Shuttle arm) where there is no gravity load, links are designed to be rather slim and their elastic deformation becomes relevant.

Links flexibility becomes a more serious problem when contact tasks are at stake. Interaction of links flexibility and environment elastodynamics produces new complex effects.

Some of the initial research in flexible-robot dynamics looked for a simple but rapid solution of a general flexible chain [6, 7], while other research tried to find a more exact model for some practical examples, e.g., single-link or two-link flexible arms [8, 9]. Almost all existing dynamic models for flexible multi-link arms are formed by some specific discretization methods, such as lumped mass [6, 7], assumed modes [10], or finite elements [11]. A few researchers considered the computational efficiency of proposed procedures [12, 13]. Real time calculation has recently become significant in the advanced control algorithms for flexible robots [14]. Many researchers considered the linear deformation and neglected the effects of coupling between the components of deformation. Novel research results [15, 16] take account of these problems, thus producing more general models.

At the beginning we recall the dynamic model of a rigid-body robot. Model (2.56), that is,

$$H(q)\ddot{q}+h(q,\dot{q})=\tau,$$

describes the unconstrained motion. Our idea is to discuss contact problems. In this case one may say that free motion appears in the first phase of any contact problem – approaching the contact position. From the viewpoint of mechanics, contact means a constraint imposed to the motion of the end-effector. Reactions appear in the form of a force and/or a torque, acting along restricted coordinates of the end-effector. If m DOFs are "lost" due to a constraint, m-component reaction F appears. In this case the model becomes

$$H(q)\ddot{q}+h(q,\dot{q})=\tau+J^{c^T}(q)F, \tag{4.40}$$

where Jacobian J^{c^T} $(n\times m)$ relates the constrained functional coordinates and the robot joint coordinates q. This was explained in Chapter 3 for a geometrical constraint and in Chapters 4 and 5 for soft and dynamic constraints.

Now, we introduce link elasticity and examine its effects upon the mechanism dynamics. In the discussion, we will assume the joints are completely rigid. Note that introduction of joint elasticity can be done in the same way as it was done in the previous chapters (e.g., Section 2.3 in Ch. 2.).

During the deformation all points of a link move relative to their undeformed locations, changing by that the relative positions, i.e. mutual distances and angles (Fig. 4.43). The figure represents the elastic displacement of a robot link (L_i) due to deformation of previous links and its own relative displacement with respect to the undeformed shape (L_i^0). Let the position of a point (marked by M) be defined by the vector $\vec{R}_i$ in the global frame B_i, and by $\vec{r}_i$ in the local frame B_i^0 (attached to each link).

Position vectors now depend not only on time (as is the case for rigid-body systems), but also on the position of the point M on the mechanism link. So, it holds that

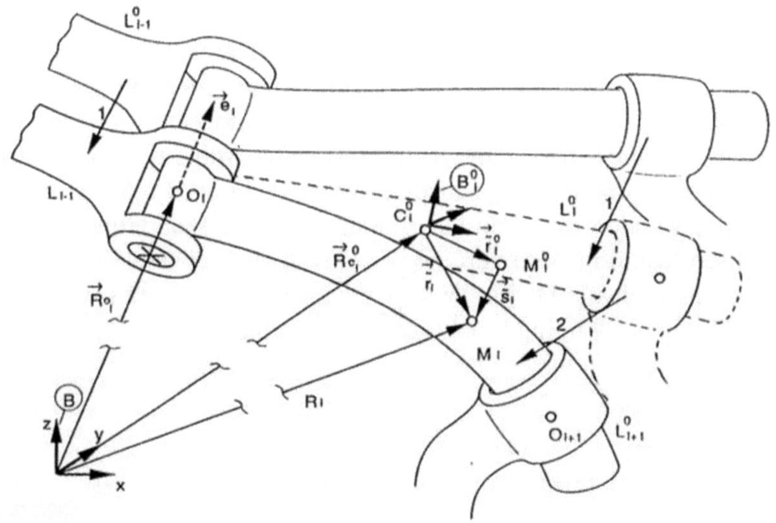

Figure 4.43. A robot link after deformation.

$$\vec{\bar{r}}_i = \vec{\bar{r}}_i^{\,0} + \vec{\bar{s}}_i \tag{4.41}$$

where $\vec{\bar{r}}_i^{\,0} = (\bar{x}_i^0, \bar{y}_i^0, \bar{z}_i^0)$ is the position vector of point M before the deformation, and $\vec{\bar{s}}_i$ is the displacement. The lines above the letters indicate that the vectors are expressed in the link-fixed frame. In the classical elasticity theory, the elastic deformation vector is a continuous function of local coordinates and time: $\vec{\bar{s}}_i = \vec{\bar{s}}_i(\bar{x}_i^0, \bar{y}_i^0, \bar{z}_i^0; t)$. Hence, the system possesses an infinite number of DOFs. As is known, however, for practical reasons the elastic deformations of complex mechanical systems are usually considered using approximation methods of structural dynamics. These methods transfer the dynamic modeling into the domain of discrete analysis, assuming small deformations and expanding the displacement vector in a series:

$$\begin{aligned}\vec{\bar{s}}_i(\bar{x}_i^0, \bar{y}_i^0, \bar{z}_i^0; t) = &\sum_{\alpha=1}^{N_i} \vec{\bar{f}}_i^{\,\alpha}(\bar{x}_i^0, \bar{y}_i^0, \bar{z}_i^0)\theta_i^\alpha(t) + \\ &\frac{1}{2}\sum_{\alpha=1}^{N_i}\sum_{\beta=1}^{N_i} \vec{\bar{f}}_i^{\,\alpha\beta}(\bar{x}_i^0, \bar{y}_i^0, \bar{z}_i^0)\theta_i^\alpha(t)\theta_i^\beta(t) + \quad \ldots\end{aligned} \tag{4.42}$$

where the time-variable $N_i \times 1$ vector

$$\Theta_i = [\theta_i^1 \dots \theta_i^\alpha \dots \theta_i^{N_i}]^T \tag{4.43}$$

is called the vector of generalized coordinates of relative flexible motion of a link. In a mathematical sense, several approximation methods of elastodynamic analysis (e.g., lumped masses, finite elements, or assumed modes) can be represented in the form (4.42). According to this, elements of vector Θ_i can represent elastic displacements of lumped masses, motion of finite-elements nodes, or amplitude of assumed-mode shape functions.

The discretization (4.42) defines an equivalent mechanical system with a finite number of elastic DOFs: $N = \sum_n N_i$, where summation is made over all links. The corresponding block-vector of coordinates is:

$$\Theta(t) = [\Theta_1 \dots \Theta_n]^T . \tag{4.44}$$

Hence, in a general form the nonlinear dynamic model of a flexible robot can be written as [15, 16]:

$$\begin{aligned} &H_{qq}(q,\Theta)\ddot{q} + H_{q\Theta}(q,\Theta)\ddot{\Theta} + h_q(q,\dot{q},\Theta,\dot{\Theta}) + \\ &\qquad\qquad + \Theta^T K_{q\Theta}\Theta = \tau + J_q^{cT}(q,\Theta)F \\ &H_{q\Theta}(q,\Theta)\ddot{q} + H_{\Theta\Theta}(q,\Theta)\ddot{\Theta} + h_\Theta(q,\dot{q},\Theta,\dot{\Theta}) + \\ &\qquad\qquad + K_{\Theta\Theta}\Theta = J_\Theta^{cT}(q,\Theta)F \end{aligned} \tag{4.45}$$

where:

$H_{qq}, H_{q\Theta}, H_{\Theta\Theta}$	– submatrices of flexible-robot inertia matrix,
h_q, h_Θ	– vectors encompassing the Coriolis', centrifugal, and gravitational effects,
$K_{q\Theta}$	– stiffness matrix describing nonlinear deformation effects (e.g., due to change of length of links connected with sliding joints),
$K_{\Theta\Theta}$	– block-diagonal structural stiffness which involves stiffness matrices of all links,
J_q^c, J_Θ^c	– Jacobian submatrices mapping generalized joint and elastic velocities into end-effector velocities.

Now, one can compare the form of dynamic model of a rigid-body robot (eq. (4.40)) and the one of a flexible robot (eq. (4.45)). The presented analysis shows that the introduction of link flexibility does not change the methodology of describing the contact dynamic effects. Contact is still modeled through the reaction vector F in dynamic model (4.45). However, the results, i.e. the system behavior, might be rather different. Additionally, if contact is followed by a deformation of the environment, and if the robot is flexible, one may expect a superposition of oscillations with all the complexity that it introduces.

Following the defined methodology for analysis of contacts, it is possible to implement all different models of environment: rigid contact, soft contact, and elestodynamic contact. It is done by using the appropriate form of contact force (as in Chapter 3 and the preceding sections of this chapter).

Robot with flexible links and flexible joints

Now it would be interesting to see how both robot-flexibility effects, joints and links, would "work" together. One should note that models (4.40) and (4.45), describing the rigid-link and the flexible-link robot respectively, involve joint torques in the form of an n-dimensional vector τ. Torques apply to joints shafts. This opens the possibility of introducing deformation in the transmission system between a joint shaft and the corresponding motor. Section 2.3. elaborated this issue for one example of actuator and transmission – DC motor followed by Harmonic-Drive which features torsion effects.

The first two equations from (4.29) defined the dynamics of a DC motor actuating joint i. Motor motion was described by the shaft angle θ_i. For a nondeformable transmission this angle is directly dependent on the joint-shaft angle q_i: $q_i = \theta_i / N_i$, where N_i is the transmission ratio. In the case of a flexible tranasmission, angles θ_i and q_i are independent, thus increasing the overall number of DOF. The joint torque now has the elastic nature and is expressed by the third equation from (4.29). So, model (4.45) for the flexible-link mechanism, along with expressions (4.29), $i = 1,...,n$, for the actuators and deformable transmissions, describe the overall robot dynamics.

REFERENCES

1. A. De Luca, C. Manes, "Hybrid Force/Position Control for Robots in Contact with Dynamic Environment", Proc. of SYROCO 91, Vienna, pp. 377-382, 1991.

2. M. H. Raibert, J. J. Craig, "Hybrid Position/Force Control of Manipulators", Journal of Dynamic Systems, Measurements and Control, Vol. 103, pp. 126-133, 1981.
3. M. Vukobratovic, Y. Ekalo, "New Approach to Control of Robotic Manipulators Interacting with Dynamic Environment", Robotica, Vol. 14, No. 1, pp. 31-39, 1996.
4. Y. Ekalo, M. Vukobratovic, "Quality of Stabilization of Robot Interacting with Dynamic Environment", Journal of Intelligent and Robotic Systems, Vol. 14, pp. 155-179, 1995.
5. M. Vukobratovic, R. Stojic, Y. Ekalo, "Contribution to the Problem Solution of Position/Force Control of Manipulation Robots in Contact with Dynamic Environment – A Generalization", IFAC Automatica, Vol. 34, No. 10, 1998.
6. M. Vukobratovic, V. Potkonjak, Dynamics of Manipulation Robots: Theory and Application, Springer-Verlag, 1982.
7. M. Vukobratovic, V. Potkonjak, Applied Dynamics and CAD of Manipulation Robots, Springer-Verlag, 1985.
8. R. H. Cannon, E. Schmitz, "Initial Experiments on the End-Point Control of a Flexible One-Link Robot", Journal of Robotics Research 3, pp. 62-75, 1984.
9. T. Fukuda, A. Arakawa, "Modeling and Control Characteristics for a Two-Degrees-of-Freedom Coupling System of Flexible Robotic Arm", JSME 30, pp. 1458-1464, 1987.
10. A. Truckenbrodt, "Dynamics and Control Methods for Moving Flexible Structures and Their Application to Industrial Robots", Proc of 5th World Congress on Theory of Machines and Mechanisms, publ. ASME, 1981.
11. H. W. Sunada, Dynamic Analysis of Flexible Spatial Mechanisms and Robotic Manipulators, Ph. D. Thesis, Univ. of California, Los Angeles, 1981.
12. W. J. Book, "Recursive Lagrangian Dynamics of Flexible Manipulator Arms", Journal of Robotics Research, 3, pp. 87-101, 1984.
13. J. O. King, V. G. Gourishankar, R. E. Rink, "Lagrangian Dynamics of Flexible Manipulators Using Angular Velocities Instead of Transformation Matrices", IEEE Trans. on SMC, 17, pp. 1059-1068, 1987.
14. M. Uchiyama, A. Conno, "Computed Acceleration Control for the Vibration Suppression of Flexible Robotic Manipulators", Proc. 5th ICAR, Pisa, Italy, pp. 126-131, 1991.
15. D. Surdilovic, M. Vukobratovic, "One Method for Efficient Dynamic Modeling of Flexible Manipulators", Mechanisms and Machine Theory, Vol. 31, No. 3, pp. 397-315, 1996.
16. D. Surdilovic, M. Vukobratovic, "Deflection Compensation for Large Flexible Manipulators", Mechanisms and Machine Theory, Vol. 31, No. 3, pp. 317-329, 1996.

Chapter 5

ROBOT INTERACTING WITH COMPLEX DYNAMIC ENVIRONMENT

The discussion about contact systems conducted in the previous chapters was based on one common idea – analysis of the robot with the constraints imposed to the motion of its end-effector. The constraints represented the robot environment. The discussion started with geometric constraints (Chapter 3). As a suitable example, the surface-typc constraint was elaborated in detail. After that, elasticity in the contact zone was introduced: first, the deformation through stiffness and damping, and then elastodynamics through lumped masses. Although the problems discussed were rather complex, they could be simply described as "the robot's end-effector in contact with the support". It should be pointed out that many practical problems may be treated in this way (writing, assembly tasks, deburring, and other process operations, etc.). The restriction was that the object processed was placed on an immobile support or, if mobile, it moved according to a prescribed law that could not be affected by the robot action.

We are now going to broaden the discussion by considering a more complex environment. Let us start with a few examples. Suppose that the robot-processed object is placed on a positioner that is servodriven. In this case the positioner (along with the object) represents the environment, and such environment cannot be treated as a geometric constraint. The introduction of deformation in the contact zone is generally desired but cannot help with the problem faced. The problem arises from the fact that the robot action may affect the motion of the positioner. In the reverse direction – motion of the positioner affects the robot. In fact, we now have two mechanical systems and the contact between them. A very similar problem appears if the object is placed on some transport system. In a more complex case another robot may replace the positioner. Then, we face the problem of bilateral manipulation. Contact effects are still very important. In order to make the discussion more general we shall consider two mechanical systems being in contact and treat them in an equal way.

The next generalization will follow from the more complex treatment of contact deformation. We shall take into account the deformation on both sides of contact, on the robot side and on the environment side.

1. MATHEMATICAL FORMULATION OF RIGID-BODY CONTACT

Let us start our discussion of contact dynamics by considering two rigid-body systems. The systems are assumed to be in the form of active chains (Fig. 5.1). The intention is to make the discussion general and treat the two systems equally. However, we find that for an easier understanding it is useful to name the systems. Let one of them be termed the *working system* (or *robot*) and the other *complex dynamic environment* (or simply *environment*). The discussion about two mechanical systems in contact need not be related to robotics. However, we are mainly interested in the robotic implementation. Thus, under the term working system we mainly understand some kind of manipulator (a simple manipulation system or a robot). What does the environment mean? It is a dynamic system with active or passive DOFs working in cooperation with a robot or assisting it. In practical problems, it can represent a rotating or coordinate table, some more complex positioner, conveyor, transport cart, CNC machine (e.g., lathe), another robot, etc. The names introduced for the two systems do not restrict the possibility of applying the following discussion to other types of mechanical systems.

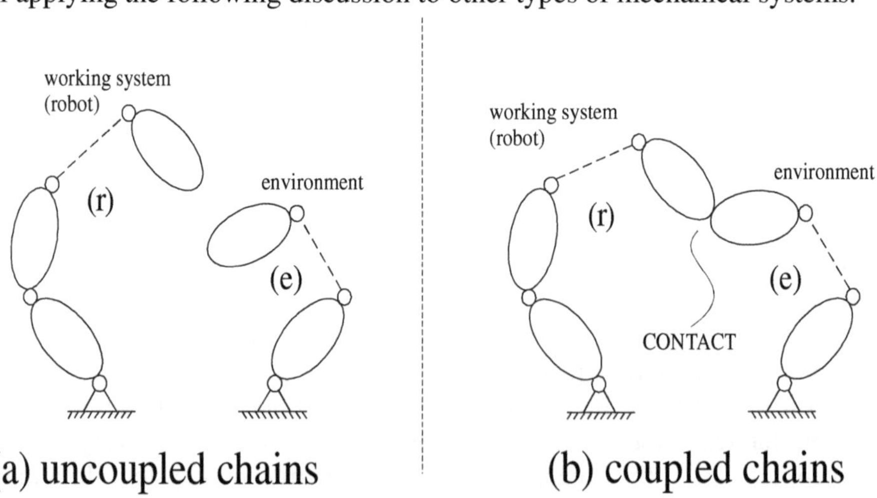

Figure 5.1. Two kinematic chains: (a) – in free motion , and (b) – in contact motion

Initially, we consider the two geometrically and dynamically uncoupled chains (Fig. 5.1a) and write their dynamic models. Let the working system (robot) have n_r DOFs (index "r" will be used for all working system variables). The environment has n_e DOFs (index "e" will be used for all environment variables). If the chain (r) (see Fig. 5.1) is a robot, then it usually has $n_r = 6$ DOFs. However, for the environment chain (e) one can hardly predict the number of DOFs. If a rotating table is in question, then it has $n_e = 1$ DOF. For an *X-Y* table, it would be $n_e = 2$, and if another robot is in question then it might be $n_e = 6$.

The internal position vectors are $q_r = (q_{r1}, \dots, q_{rn_r})$ and $q_e = (q_{e1}, \dots, q_{en_e})$. They define the motion of joints. The position of the entire system may be expressed by the united vector of dimension $n = n_r + n_e$:

$$q = \begin{bmatrix} q_e \\ q_r \end{bmatrix} \tag{5.1}$$

The dynamics of the two decoupled chains can be described by the model (see Chapter 2):

$$H_e(q_e)\ddot{q}_e + h_e(q_e, \dot{q}_e) = \tau_e \tag{5.2}$$

$$H_r(q_r)\ddot{q}_r + h_r(q_r, \dot{q}_r) = \tau_r \tag{5.3}$$

where τ_e and τ_r are the joint torque vectors. There are reasons why the environment model is placed first, before the robot model. This will be clear after introducing the set of functional coordinates.

For each chain, one may introduce a set of functional coordinates describing the position in the most appropriate way for a particular problem. Let $s_e = (s_{e1}, \dots, s_{en_e})$ and $s_r = (s_{r1}, \dots, s_{rn_r})$ be the functional coordinates for the two chains. Relation between the coordinates q and s represents the kinematic model. For the environment, s-coordinates are defined as absolute. This means that these coordinates describe the environment position with respect to the immobile (absolute) frame. Very often the set s_e coincides with q_e since it may be impossible to find a more suitable set of coordinates. In a general case,

$$s_e = s_e(q_e) \tag{5.4}$$

represents a nonlinear function. The robot's operation assumes some prescribed relation with the environment. The relation includes kinematic behavior and sometimes also dynamic effects. Hence, it is suitable to define the s-coordinates of a robot as relative. They describe the position of the robot (its end-effector) with respect to the environment (i.e. a frame fixed to the terminal link of the environment chain (e)). Accordingly, the transformation of the coordinates has the form:

$$s_r = s_r(q_e, q_r). \tag{5.5}$$

The set of functional coordinates of the overall system is:

$$s = \begin{bmatrix} s_e \\ s_r \end{bmatrix}. \tag{5.6}$$

Let us now discuss the contact between the two chains (Fig. 5.1b). How do we express the contact mathematically? Introduction of the robot's functional coordinates describing the relative position with respect to the environment, offers the possibility of applying the same approach as applied in the previous chapters. We note that the relative coordinates s_r are defined so as to support the prescribed task. Hence, it is justified to assume that the contact can be described by means of zero-values of some of these coordinates. If we suppose that the contact restricts m DOFs, then m coordinates from the set s_r reduce to zero. Now it is convenient to partition the vector s into two subvectors but in a way different from equation (5.6). Let the set s^c contain the coordinates constrained by the contact (all from s_r) and have dimension m ($m \le n_r$). The remaining coordinates from s_r ($n_r - m$ of them) are free. If s^f is the set of free coordinates of the entire system, then it includes the complete vector s_e, plus $n_r - m$ elements of the vector s_r. Dimension of s^f is $n_e + n_r - m$. The new partition is

$$s = \begin{bmatrix} s^f \\ s^c \end{bmatrix}. \tag{5.7}$$

The constraint (contact) is now expressed as

$$s^c = s^c(q_e, q_r) = 0 . \tag{5.8}$$

This means that the rigid-body contact is considered and modeled by applying geometric constraints. From the dynamic point of view, contact causes a reaction force (contact force). The number of force components equals the number of restricted DOFs. In this case the model (5.2), (5.3) should include the contact force F of dimension m:

$$H_e(q_e)\ddot{q}_e + h_e(q_e,\dot{q}_e) = \tau_e + J_e^{c^T}(q_e,q_r)F \tag{5.9}$$

$$H_r(q_r)\ddot{q}_r + h_r(q_r,\dot{q}_r) = \tau_r + J_r^{c^T}(q_e,q_r)F \ , \tag{5.10}$$

where J_e^c and J_r^c are Jacobian matrices: $J_e^c = \partial s^c / \partial q_e$ and $J_r^c = \partial s^c / \partial q_r$, of dimensions $m \times n_e$ and $m \times n_r$, respectively.

Because of the constraint (5.8), the coordinates q_e and q_r appearing in (5.9) and (5.10) are not independent. Accordingly, the models (5.9) and (5.10), along with the condition (5.8), describe the dynamics of the overall system. It is more convenient if the constraint relation is written in a differential form. The first derivative of s^c gives:

$$\dot{s}^c = \frac{\partial s^c}{\partial q_e}\dot{q}_e + \frac{\partial s^c}{\partial q_r}\dot{q}_r = J_e^c \dot{q}_e + J_r^c \dot{q}_r \ , \tag{5.11}$$

and the second derivative:

$$\ddot{s}^c = J_e^c \ddot{q}_e + J_r^c \ddot{q}_r + A^c \ , \tag{5.12}$$

where the matrices J_e^c and J_r^c are the same ones as introduced in (5.9) and (5.10) and the adjoint vector is

$$A^c = (\partial^2 s^c)/(\partial q_e)^2 \, \dot{q}_e^2 + (\partial^2 s^c)/(\partial q_e \partial q_r) \, \dot{q}_e \dot{q}_r + \\ + (\partial^2 s^c)/(\partial q_r \partial q_e) \, \dot{q}_r \dot{q}_e + (\partial^2 s^c)/(\partial q_r)^2 \, \dot{q}_r^2 \ .$$

The constraint (5.8), written in a second-order differential form, becomes

$$\ddot{s}^c = J_e^c \ddot{q}_e + J_r^c \ddot{q}_r + A^c = 0 \, . \tag{5.13}$$

The dynamics is now described by the model (5.9), (5.10), along with the relation (5.13). This is a set of $n_e + n_r + m$ scalar equations that can be solved for the same number of scalar unknowns: n_e components of acceleration $\ddot{q}_e$, n_r components of acceleration $\ddot{q}_r$, and m components of the force F. Thus, simulation is possible.

2. MODELING FRICTION EFFECTS

In the approach used, the contact influenced the subset of functional coordinates: s^c was kept zero by means of the reaction force F acting in the direction of the coordinates s^c. The remaining coordinates (the subset s^f) were free and there was no resistance in these directions. However, in a real situation, some kind of friction force always exists. Friction appears between two bodies in contact. We are interested in the friction related to the contact of the two chains, (*r*) and (*e*). We remember that the coordinates s_r defined the position of the terminal link of the robot chain (*r*) with respect to the terminal link of the environment chain (*e*). Some of these coordinates were restricted by the contact (the set s^c of dimension *m*) while the others were free and belonged to the set s^f (together with all the coordinates from s_e, see (5.6) and (5.7)). If those coordinates from s_r that are free form a subset s_r^f of dimension $n_r - m$, then $s^f = (s_e, s_r^f)$. Generally, each of the coordinates from s_r^f may be subject to friction. Such discussion of friction would be rather complex. It would involve details about the constructive realization of the contact. In order to avoid an overly complex elaboration, we confine our consideration to *point-to-surface contact*. This means that a particular point of the robot terminal link (we call this point "the tip") is in contact with a surface that belongs to the terminal link of the environment. A very similar task, called the writing task, was considered in the previous chapters. In this problem the number of restricted coordinates is $m = 1$ and the reaction force F between the two terminal links has one component only (see Fig. 5.2). The vector of relative motion (s_r) has $n_r - 1$ free components. Two of these free coordinates (the ones describing the tip motion along the surface) are subject to friction (Fig. 5.2).

The discussion that follows is rather similar to that in Chapters 3 and 4 (see Section 2. in Ch. 3 and Section 1.2 in Ch. 4). The difference is that now the robot writes over a support that has its own dynamics. Friction force vector is opposite to the relative velocity and has the value μF. In Chapter 3, the coordinates subject to friction were the first two components of vector s and at the same time of the vector s^f. By analogy, they are now the first two components of s_r and of s_r^f. Let s_{r1} and s_{r2} be these two coordinates and let them form the subvector $s_r^{(1,2)} = (s_{r1}, s_{r2})$. Their derivatives, $\dot{s}_{r1}$

and $\dot{s}_{r2}$, define the relative velocity. The derivatives may be expressed in terms of joint velocities:

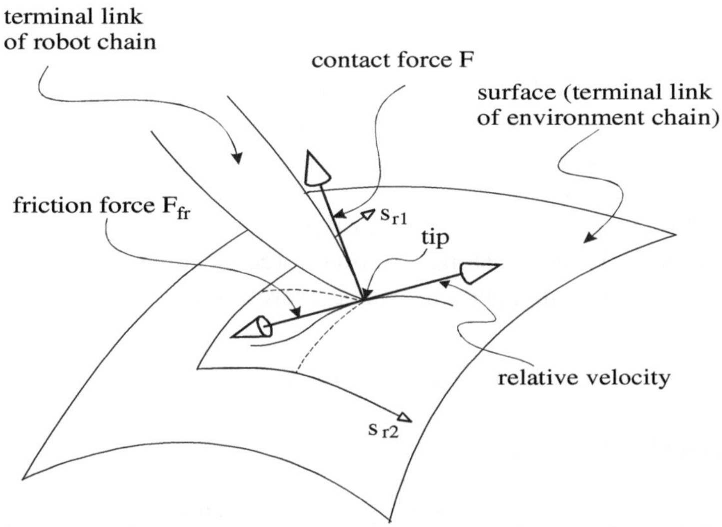

Figure 5.2. Friction in the case of point-to-surface contact. Friction force appears in the direction of the coordinates that describe the relative motion of the tip. Friction vector is opposite to the relative velocity.

$$\dot{s}_r^{(1,2)} = \begin{bmatrix} \dot{s}_{r1} \\ \dot{s}_{r2} \end{bmatrix} = J_{rr}^{(1,2)}(q_e, q_r)\,\dot{q}_r + J_{re}^{(1,2)}(q_e, q_r)\,\dot{q}_e\,, \tag{5.14}$$

where $J_{rr}^{(1,2)} = \partial s_r^{(1,2)} / \partial q_r$ and $J_{re}^{(1,2)} = \partial s_r^{(1,2)} / \partial q_e$. The unit vector of relative velocity (projections onto the axes s_{r1} and s_{r2}) is: $\dot{s}_r^{(1,2)} / \left(\dot{s}_{r1}^2 + \dot{s}_{r2}^2\right)^{1/2}$. Friction force, being opposite to the relative velocity, should also be expressed by its two components along s_{r1} and s_{r2}:

$$F_{fr} = -\mu F\, \dot{s}_r^{(1,2)} / \left(\dot{s}_{r1}^2 + \dot{s}_{r2}^2\right)^{1/2}. \tag{5.15}$$

This form of friction force now enters the mathematical model of system dynamics. Equations (5.9) and (5.10) become:

$$\begin{aligned} H_e(q_e)\,\ddot{q}_e + h_e(q_e, \dot{q}_e) = \tau_e + J_e^{cT}(q_e, q_r)F + \\ + J_{re}^{(1,2)T}(q_e, q_r)F_{fr} \end{aligned} \tag{5.16}$$

$$H_r(q_r)\ddot{q}_r + h_r(q_r,\dot{q}_r) = \tau_r + J_r^{c^T}(q_e,q_r)F + \\ + J_{rr}^{(1,2)^T}(q_e,q_r)F_{fr} \tag{5.17}$$

With the friction expressed by (5.15), the dynamic model (5.16), (5.17) along with relation (5.13) (describing the contact), can be solved for $n_e + n_r + 1$ scalar unknowns: n_e components of the acceleration $\ddot{q}_e$, n_r components of the acceleration $\ddot{q}_r$, and one component of the force F. Thus, simulation is possible.

We should remember that the above discussion holds while the tip slides over the surface. When it stops, friction force becomes unknown. This situation is called jamming, and this can be expressed by:

$$\dot{s}_r^{(1,2)} = J_{rr}^{(1,2)}(q_e,q_r)\,\dot{q}_r + J_{re}^{(1,2)}(q_e,q_r)\,\dot{q}_e = 0, \tag{5.18}$$

or in the second-order form:

$$\ddot{s}_r^{(1,2)} = J_{rr}^{(1,2)}(q_e,q_r)\,\ddot{q}_r + J_{re}^{(1,2)}(q_e,q_r)\,\ddot{q}_e + A_r^{(1,2)} = 0 \tag{5.19}$$

The dynamic model (5.16), (5.17) along with relations (5.13) (describing the contact) and (5.19) (describing jamming), consists of $n_e + n_r + 1 + 2$ scalar equations that can be solved for the same number of unknowns: $\ddot{q}_e$ (n_e components), $\ddot{q}_r$ (n_r components), F (one component), and F_{fr} (two components). Jamming ends and the tip starts to move when the magnitude of the friction force reaches the value μF.

3. INTRODUCING CONTACT ELASTODYNAMICS

In the previous discussion we assumed that both chains consisted of rigid bodies. We now introduce elastic effects in the contact zone. Since the idea is to treat the two chains equally, it is assumed that deformation appears on both sides of the contact (on the robot side and environment side). This is the difference between the discussion given in Chapter 4, where the robot end-effector was considered rigid and deformation existed on the environment side only.

Since deformation is present, the contact cannot be considered as a geometric constraint. Coordinates from the set s^c need no longer be zero. In fact, these coordinates represent the deformation. From the standpoint of geometry and kinematics, they are free and may take arbitrary values.

However, they are subject to strong contact forces that make their values small. Hence, we still talk about constrained coordinates.

Let us discuss the kinematic expression (5.12) that relates the coordinates s^c and the internal coordinates q_e, q_r. This expression calculates some position variables of the robot tip. The expression includes only "rigid parameters" and does not take care of the deformation which appears in the contact zone on the robot side. Thus, the calculated position of the tip is "fictitious". It shows where the tip would be if the end-effector was rigid. Let us explain this. If there were no deformation at any side of contact (both the environment and end-effector are rigid) the coordinates s^c would be zero (Fig. 5.3a). If deformation appeared on the environment side only, the end-effector would move "downwardly", deforming the environment (Fig. 5.3b), and the coordinates s^c would take negative values. The magnitudes of s^c would represent the environment deformation. We now also introduce deformation on the robot side. In this case, we have to distinguish between the position of the end-effector and the position of its tip (Fig. 5.3c). Deformation on the robot side appears in the small contact zone. Hence, we shall notice that the end-effector performs additional movement "downwardly" although its tip (real tip) does not move. The additional motion is due to the deformation of the end-effector in the tip zone. The coordinates s^c define the "fictitious" tip (as if there were no deformation on the robot side) and they will increase in the negative direction. Thus, s^c represent the sum of deformations on both sides of contact.

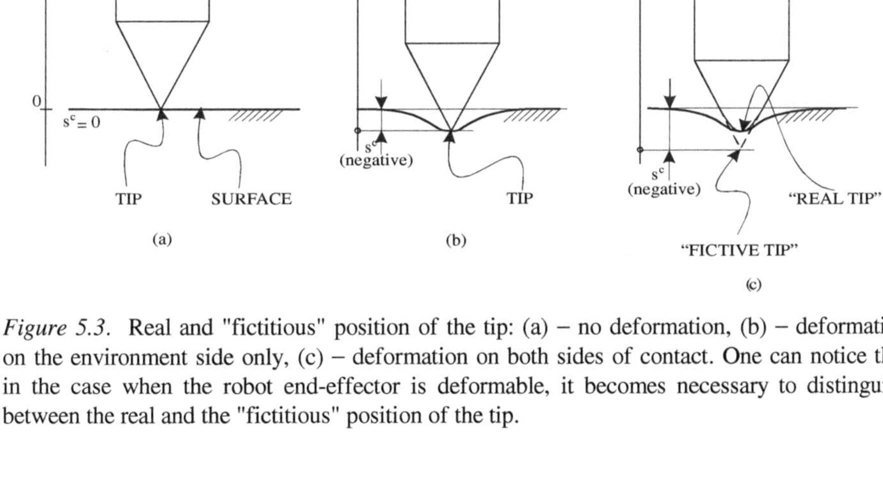

Figure 5.3. Real and "fictitious" position of the tip: (a) – no deformation, (b) – deformation on the environment side only, (c) – deformation on both sides of contact. One can notice that in the case when the robot end-effector is deformable, it becomes necessary to distinguish between the real and the "fictitious" position of the tip.

Deformation will be treated using the lumped masses approach. The same approach was used in Section 2 in Ch. 4. Here we rely on the assumption that deformations are small and deviations along different directions are mutually independent.

On each side of the contact we introduce external and internal deformation, and the corresponding equivalent masses (two-stage deformation model). Let us show this for the coordinate s_i^c (from the set s^c) which represents one direction of contact (Fig. 5.4). On the robot side, external deformation is marked by s_{ri}^{c1} and internal deformation by s_{ri}^{c2}. Equivalent masses are $m_{ri}^{(e)1}$ and $m_{ri}^{(e)2}$. On the environment side, external and internal deformations are s_{ei}^{c1} and s_{ei}^{c2}, respectively and the equivalent masses are $m_{ei}^{(e)1}$ and $m_{ei}^{(e)2}$. Due to the nature of the contact, all deformations are negative. The coordinate s_i^c is the sum of these four deformations:

$$s_i^c = s_{ri}^{c2} + s_{ri}^{c1} + s_{ei}^{c1} + s_{ei}^{c2} \ . \tag{5.20}$$

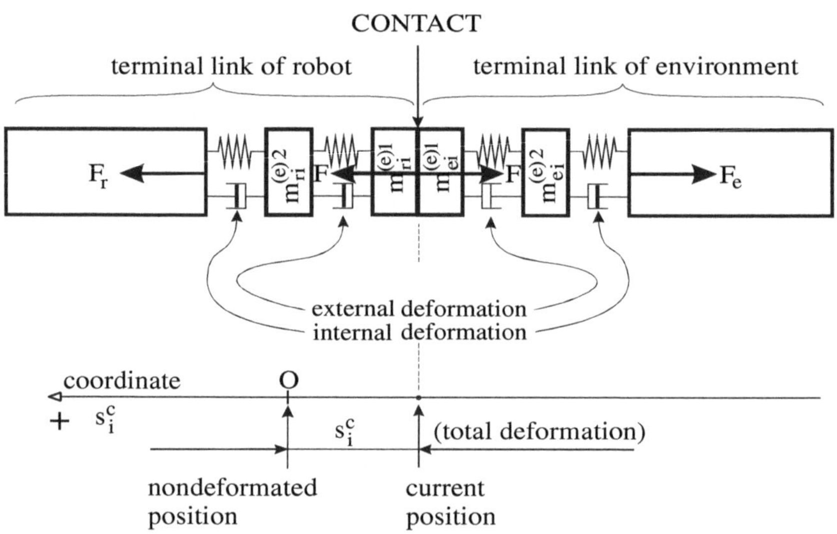

Figure 5.4. Model of contact: multibody lumped system

For all coordinates from the set s^c one may write:

$$s^c = s_r^{c2} + s_r^{c1} + s_e^{c1} + s_e^{c2} \ , \tag{5.21}$$

where all variables are m-dimensional vectors (such as, vector $s_r^{c2} = [s_{r1}^{c2} \ s_{r2}^{c2} \ \ldots \ s_{rm}^{c2}]^T$).

Now, we should recall that the dynamics of rigid-body-contact system was described by equations (5.9) and (5.10), along with the relation (5.13) expressing the constraint. For the point-to-surface contact and the presence of friction, equations (5.16) and (5.17) replace (5.9) and (5.10). It is clear from Fig. 5.4 that equivalent masses are "taken" from the terminal links of the two chains. If the robot chain masses are $m_{r1}, m_{r2}, \ldots, m_{rn_r}$, then the terminal link mass m_{rn_r} includes the equivalent masses $m_{ri}^{(e)1}$ and $m_{ri}^{(e)2}$, $i = 1,\ldots,m$. Analogous conclusion holds for the environment. Looking at the model of contact presented in Fig. 5.4, one finds that equations (5.9), (5.10), and (5.16), (5.17) do not hold in the case of contact with deformation. After extracting the equivalent masses of the terminal links, the forces acting upon the two chains become different: F_r will act upon the robot, and F_e upon the environment. They are both different from the contact force F. Mathematical models will now change. Equations (5.9) and (5.10) will become:

$$H_e(q_e)\ddot{q}_e + h_e(q_e,\dot{q}_e) = \tau_e + J_e^{c^T}(q_e,q_r)F_e \tag{5.22}$$

$$H_r(q_r)\ddot{q}_r + h_r(q_r,\dot{q}_r) = \tau_r + J_r^{c^T}(q_e,q_r)F_r\ , \tag{5.23}$$

and equations (5.16) and (5.17) will be:

$$\begin{aligned} H_e(q_e)\ddot{q}_e + h_e(q_e,\dot{q}_e) = \tau_e + J_e^{c^T}(q_e,q_r)F_e + \\ + J_{re}^{(1,2)^T}(q_e,q_r)F_{fr} \end{aligned} \tag{5.24}$$

$$\begin{aligned} H_r(q_r)\ddot{q}_r + h_r(q_r,\dot{q}_r) = \tau_r + J_r^{c^T}(q_e,q_r)F_r + \\ + J_{rr}^{(1,2)^T}(q_e,q_r)F_{fr} \end{aligned} \tag{5.25}$$

Note that the friction force still depends on the force F (see expression 5.15) and not on F_e and F_r.

At this point some explanations should be given. Strictly speaking, if equivalent masses are extracted from the terminal links, then the chain inertia is changed and the coefficients *H* and *h* in the models (5.22)–(5.25) should be recalculated. However, we consider the motion of equivalent masses as consisting of two components: transport motion that follows from the motion of joints, and local (micro) motion that is due to the deformation. In mathematical modeling, equations of chain dynamics will cover the motion of the chain and also the transport motion of equivalent masses. For this reason we use the complete masses of the terminal links when forming the models (5.22)–(5.25). Coefficients H and h will be the same as they were in (5.9), (5.10), (5.16), and (5.17). The dynamics of local motion of equivalent masses will be discussed separately.

Without losing generality, we concentrate on the model (5.22), (5.23) and avoid the elaboration of friction. The model consists of $n_e + n_r$ scalar equations that should be solved for $n_e + n_r + 2m$ scalar unknowns: n_e components of the acceleration $\ddot{q}_e$, n_r components of the acceleration $\ddot{q}_r$, m components of the force F_e, and m components of the force F_r. Thus, some equations are missing.

The geometric constraint (5.13) does not hold and cannot be used to support the determination of the forces F_e and F_r. Expressions for the forces can be obtained from the analysis of deformation. Force F_e can be found on the basis of the internal deformation on the environment side. For the direction s_i^c, the force component is

$$F_{ei} = -k_{ei}^2 s_{ei}^{c2} - b_{ei}^2 \dot{s}_{ei}^{c2}, \tag{5.26}$$

where the coefficients k_{ei}^2 and b_{ei}^2 characterize the stiffness and damping of the internal deformation on the environment side. On the robot side it holds that:

$$F_{ri} = -k_{ri}^2 s_{ri}^{c2} - b_{ri}^2 \dot{s}_{ri}^{c2}. \tag{5.27}$$

If this is applied to all directions from the set s^c (that is, $i = 1, \ldots, m$), one obtains the force vectors:

$$F_e = -k_e^2 s_e^{c2} - b_e^2 \dot{s}_e^{c2}, \tag{5.28}$$

$$F_r = -k_r^2 s_r^{c2} - b_r^2 \dot{s}_r^{c2}, \tag{5.29}$$

where the coefficients are diagonal matrices (e.g., $k_e^2 = diag\,[k_{ei}^2]$). In this way the forces F_e and F_r are eliminated from the set of unknowns. However, the deformations s_e^{c2} and s_r^{c2} are introduced. Thus, the number of unknowns in the model (5.22), (5.23) remains the same (that is $n_e + n_r + 2m$).

We now analyze the dynamics of equivalent masses (elastodynamics). It can be described applying Newton's law:

mass $\times$ acceleration = sum of forces.

For a spatial motion of a particle, the law involves absolute acceleration. In the problem that we consider here, it is more convenient to use relative acceleration since it can be expressed in terms of second derivatives of the external and internal deformation. If relative acceleration appears on the left-hand side of Newton's law then the right-hand side must include the inertial forces of the transport and Coriolis' accelerations. However, we assumed that in the absence of contact, no deformation occurred. This meant that the Coriolis' and transport inertia were in equilibrium with gravity and connection forces between particles. The contact force disturbs this equilibrium and produces relative motion of particles in the direction of the force. We call this motion deformation. Now, it is justified to use relative acceleration on the left-hand side of Newton's law and, at the right-hand side omit the transport and Coriolis' inertial forces, gravity term, and the connection forces except those that follow from the deformation. This analysis allows us to apply Newton's law in the way similar to the one used in Section 2.1 in Ch. 4, to derive the model (4.10). Thus, we come to two equations describing the dynamics of equivalent masses along the direction s_i^c on the environment side:

$$\begin{aligned} F_i &= -m_{ei}^{(e)1}(\ddot{s}_{ei}^{c1} + \ddot{s}_{ei}^{c2}) - k_{ei}^1 s_{ei}^{c1} - b_{ei}^1 \dot{s}_{ei}^{c1} \\ k_{ei}^1 s_{ei}^{c1} + b_{ei}^1 \dot{s}_{ei}^{c1} &= m_{ei}^{(e)2} \ddot{s}_{ei}^{c2} + k_{ei}^2 s_{ei}^{c2} + b_{ei}^2 \dot{s}_{ei}^{c2} \ , \end{aligned} \tag{5.30}$$

and another two describing the elastodynamics on the robot side:

$$\begin{aligned} F_i &= m_{ri}^{(e)1}(\ddot{s}_{ri}^{c1} + \ddot{s}_{ri}^{c2}) - k_{ri}^1 s_{ri}^{c1} - b_{ri}^1 \dot{s}_{ri}^{c1} \\ k_{ri}^1 s_{ri}^{c1} + b_{ri}^1 \dot{s}_{ri}^{c1} &= -m_{ri}^{(e)2} \ddot{s}_{ri}^{c2} + k_{ri}^2 s_{ri}^{c2} + b_{ri}^2 \dot{s}_{ri}^{c2} \ . \end{aligned} \tag{5.31}$$

Coefficients k_{ei}^1 and b_{ei}^1 characterize the stiffness and damping of the external deformation on the environment side, while k_{ei}^2 and b_{ei}^2 characterize the internal deformation on the same side. In an analogous way the coefficients k_{ri}^1, b_{ri}^1, k_{ri}^2, and b_{ri}^2 hold for the robot side. If (5.30) and (5.31) are generalized to cover all directions of deformation $(i = 1,..., m)$ then the matrix expressions are obtained:

$$\begin{aligned} F &= -m_e^{(e)1}(\ddot{s}_e^{c1} + \ddot{s}_e^{c2}) - k_e^1 s_e^{c1} - b_e^1 \dot{s}_e^{c1} \\ k_e^1 s_e^{c1} + b_e^1 \dot{s}_e^{c1} &= m_e^{(e)2} \ddot{s}_e^{c2} + k_e^2 s_e^{c2} + b_e^2 \dot{s}_e^{c2} \ , \end{aligned} \tag{5.32}$$

and

$$F = m_r^{(e)1}(\ddot{s}_r^{c1} + \ddot{s}_r^{c2}) - k_r^1 s_r^{c1} - b_r^1 \dot{s}_r^{c1}$$
$$k_r^1 s_r^{c1} + b_r^1 \dot{s}_r^{c1} = -m_r^{(e)2} \ddot{s}_r^{c2} + k_r^2 s_r^{c2} + b_r^2 \dot{s}_r^{c2} \quad . \tag{5.33}$$

The models of elastodynamics (5.32) and (5.33) define $4m$ scalar equations. At the same time the models introduce new unknowns: the m-component vectors F, $\ddot{s}_e^{c1}$, $\ddot{s}_e^{c2}$, $\ddot{s}_r^{c1}$, and $\ddot{s}_r^{c2}$. Thus, the model of "macro" dynamics (equations (5.22) and (5.23) with the forces F_e and F_r expressed by (5.28) and (5.29)) and the model of elastodynamics (equations (5.32) and (5.33)), if considered together, have $n_e + n_r + 4m$ scalar equations and $n_e + n_r + 5m$ unknowns: internal accelerations $\ddot{q}_e$ and $\ddot{q}_r$, force F, and deformation accelerations $\ddot{s}_e^{c1}$, $\ddot{s}_e^{c2}$, $\ddot{s}_r^{c1}$, and $\ddot{s}_r^{c2}$. Thus, m equations are still missing.

We have already concluded that the geometric constraint (5.13) does not hold here. The coordinates from the set s^c may change. Each component represents the total deformation in one direction of contact (see (5.20) and (5.21)). The deformation vector s^c is related to the q-coordinates of the two chains through the differential form (5.12). Combining (5.12) and (5.21) one obtains

$$\ddot{s}_r^{c2} + \ddot{s}_r^{c1} + \ddot{s}_e^{c1} + \ddot{s}_e^{c2} = J_e^c \ddot{q}_e + J_r^c \ddot{q}_r + A^c \, . \tag{5.34}$$

This relation, consisting of m scalar expressions, is added to the dynamic models to obtain a sufficient number of equations. To summarize, expressions (5.22), (5.23), (5.32), (5.33), and (5.34) contain $n_e + n_r + 5m$ scalar equations and can be solved for the same number of unknowns listed above. Calculation of accelerations allows simulation of the complete system dynamics.

At this point we should recall that each contact task begins with the approaching phase. In that phase the working system (robot) moves towards the environment while the environment system (e.g., a positioner) performs motion in preparation for the expected cooperation with the robot. Both chains are free. At the beginning of Section 1 we considered the free motion. Dynamics of the two chains was described by the models (5.2) and (5.3) that enable simulation. Functional coordinates were defined: s_e and s_r forming the set s according to expression (5.6). After that, a different partition of the vector s was made: subvectors s^f and s^c were introduced according to (5.7). Coordinates from the set s^c were responsible for the contact. Let us remember that in Chapters 3 and 4 we assumed that all the coordinates s^c reached zero value simultaneously at the instant of contact. We shall hold this assumption in the present discussion. So, during the approaching phase, the vector s^c is checked constantly. When it becomes zero, the contact between the two chains is made. The approaching motion is always planned so as to avoid impact (achieving zero-velocity contact). However, due to tracking errors, the velocity at the instant of contact will not be zero. This yields collision between the terminal links of the two chains and the occurrence of impact. The impact involves two equivalent masses, $m_e^{(e)1}$ and $m_r^{(e)1}$, representing the inertia of the external deformation on the two sides in contact. The impact is solved by simply applying the law of momentum conservation. After the impact, the contact phase of motion begins. The impact is usually solved under the assumption that its duration is infinitely short. However, the consequences of the impact shock are present for some time after. They are manifested as oscillations which reduce in time due to action of the control and the natural damping of the system. Hence, we talk about the transient phase before the "regular" contact motion is achieved. If the inertia of the external deformation may be neglected, then no impact occurs. However, the transient phase characterized by oscillations still exists.

4. CASE STUDY

We are going to elaborate two examples. The first one deals with contact of a robot and a transport cart. The other discusses a more complex cooperation, the problem of assembly. The peg-in-hole task is solved.

4.1 Robot in Contact with a Transport Cart

Theoretical consideration

The system is comprised of a robot working in cooperation with a transport cart. The task consists of producing a force upon an object placed on the cart. It is an in-motion operation as the cart does not stop. Force has to be applied in the horizontal and vertical directions. Fig. 5.5 shows the system configuration and explains the task.

Our aim is to apply the general model derived in the previous sections (5.1–5.3) and we now search for the correlation between the model and the system considered. The cart represents the environment (*e*). It is an active one-DOF system $(n_e = 1)$. Its drive moves it in the horizontal direction and we describe this motion by means of the coordinate $q_e = x$. The functional *s*-coordinate is defined so as to coincide with q_e, that is $s_e = q_e = x$. Inertia of the environment means the cart mass m_e (object on the cart included). The cart actuator produces the drive τ_e.

A robot with two revolute DOFs represents the working system. Now, $n_r = 2$ and the internal position vector is $q_r = (q_{r1}, q_{r2})$. The task assumes cooperation between the robot and the cart. Hence, the functional coordinates of the robot are introduced so as to define the relative position of the robot tip with respect to the cart. The frame fixed to the cart (Fig. 5.5) has the coordinates $s_r = (s_{r1}, s_{r2})$. The robot's inertia is defined by means of two masses (m_{r1} and m_{r2}) and two moments of inertia (J_{r1} and J_{r2}). Joint torques are $\tau_r = (\tau_{r1}, \tau_{r2})$.

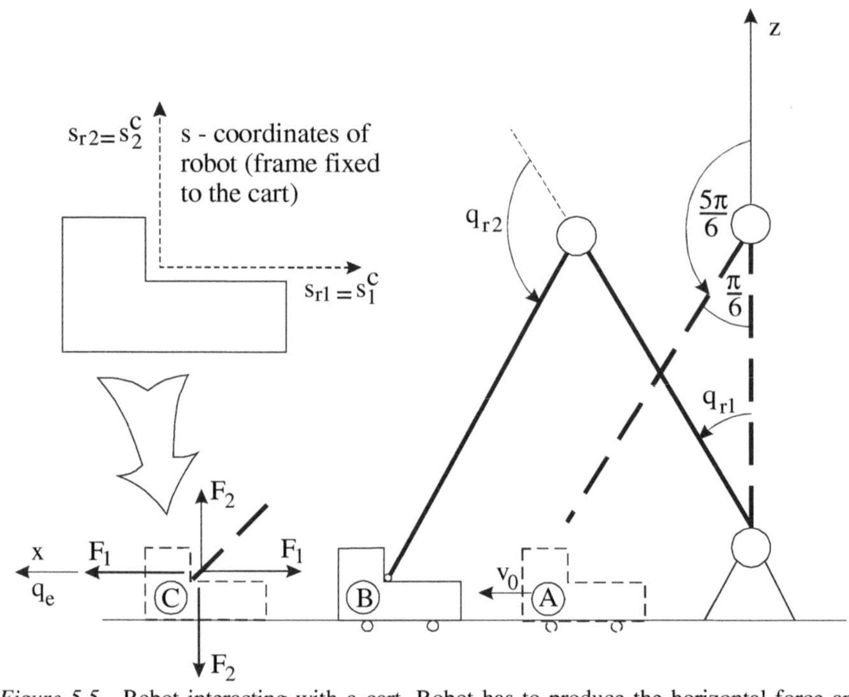

Figure 5.5. Robot interacting with a cart. Robot has to produce the horizontal force and the vertical force upon the object placed on the transport cart.

During the free motion, the robot-plus-cart system is described by the dynamic model (5.2), (5.3) that reduces to

$$m_e \ddot{q}_e = \tau_e \tag{5.35}$$

for the environment subsystem, and

$$\begin{aligned} H_{r11}\ddot{q}_{r1} + H_{r12}\ddot{q}_{r2} + h_{r1} &= \tau_{r1} \\ H_{r21}\ddot{q}_{r1} + H_{r22}\ddot{q}_{r2} + h_{r2} &= \tau_{r2} \ , \end{aligned} \tag{5.36}$$

for the robot. Coefficients H_{rij} and h_{rj} are:

$$
\begin{aligned}
&H_{r11} = J_{r1} + J_{r2} + m_{r1} l_{1C}^2 + m_{r2}(l_1^2 + l_{2C}^2 + 2 l_1 l_{2C} \cos q_{r2}) \\
&H_{r12} = H_{r21} = m_{r2} l_{2C} (l_{2C} + l_1 \cos q_{r2}) + J_{r2} \\
&H_{r22} = J_{r2} + m_{r2} l_{2C}^2 \\
&h_{r1} = -l_1 l_{2C} m_{r2} (2\dot{q}_{r2}\dot{q}_{r1} + \dot{q}_{r2}^2) \sin q_{r2} - m_{r1} g l_{1C} \sin q_{r1} - \\
&\qquad\qquad -m_{r2} g (l_1 \sin q_{r1} + l_{2C} \sin(q_{r1} + q_{r2})) \\
&h_{r2} = m_{r2} l_{2C} (l_1 \dot{q}_{r1}^2 \sin q_{r2} - g \sin(q_{r1} + q_{r2})) \quad ,
\end{aligned}
\tag{5.37}
$$

where $l_j, j = 1,2$ are the lengths of the links, and l_{jC} defines position of the MC with respect to the preceding joint (we shall adopt $l_{jC} = l_j / 2$).

We now define a mathematical relation between the q-coordinates and s-coordinates. The functional s-coordinate for the cart has already been defined so as to coincide with q:

$$s_e = q_e \ . \tag{5.38}$$

The robot's functional coordinates have been defined in Fig. 5.5 and can be expressed in terms of q_r and q_e (as indicated in expression (5.5)):

$$s_r = (s_{r1}, s_{r2}) , \tag{5.39}$$

$$
\begin{aligned}
&s_{r1} = -l_1 \sin q_{r1} - l_2 \sin(q_{r1} + q_{r2}) + q_e \\
&s_{r2} = l_1 \cos q_{r1} + l_2 \cos(q_{r1} + q_{r2}) \ .
\end{aligned}
\tag{5.40}
$$

In the phase of contact motion, the vector s should be partitioned according to (5.7). In the presented example, the contact involves the coordinates s_{r1} and s_{r2}, the coordinate s_e being free. Thus, it holds that:

$$s^f = s_e \ , \tag{5.41}$$

and

$$s^c = s_r = (s_{r1}, s_{r2}) . \tag{5.42}$$

In order to apply the dynamic model (5.22), (5.23), it is necessary to find expressions for the Jacobian matrices. Following the definition of the Jacobian matrices, one obtains:

$$J_e^c = \begin{bmatrix} 0 \\ 1 \end{bmatrix},$$
$$J_r^c = \begin{bmatrix} -l_1 \cos q_{r1} - l_2 \cos(q_{r1}+q_{r2}) & -l_2 \cos(q_{r1}+q_{r2}) \\ -l_1 \sin q_{r1} - l_2 \sin(q_{r1}+q_{r2}) & -l_2 \sin(q_{r1}+q_{r2}) \end{bmatrix}. \tag{5.43}$$

Jacobian matrices are also needed for the expression (5.13) for the case of rigid-body contact, and for expression (5.34) in the case of elastodynamic contact. These expressions also need the adjoint matrix A^c. Following the definition of the adjoint matrix, one obtains:

$$A^c = \begin{bmatrix} l_1 \sin q_{r1} \dot{q}_{r1}^2 + l_2 \sin(q_{r1}+q_{r2})(\dot{q}_{r1}+\dot{q}_{r2})^2 \\ -l_1 \cos q_{r1} \dot{q}_{r1}^2 - l_2 \cos(q_{r1}+q_{r2})(\dot{q}_{r1}+\dot{q}_{r2})^2 \end{bmatrix}. \tag{5.44}$$

The dynamic model (5.22), (5.23) can now be applied to the cart-plus-robot system. One obtains:

$$m_e \ddot{q}_e = \tau_e + F_{e1} \tag{5.45}$$

for the cart subsystem, and

$$H_{r11}\ddot{q}_{r1} + H_{r12}\ddot{q}_{r2} + h_{r1} = \tau_{r1} + J_{r11}^c F_{r1} + J_{r21}^c F_{r2}$$
$$H_{r21}\ddot{q}_{r1} + H_{r22}\ddot{q}_{r2} + h_{r2} = \tau_{r2} + J_{r12}^c F_{r1} + J_{r22}^c F_{r2} \tag{5.46}$$

for the robot. Elements of the Jacobian matrix are defined in (5.43) and the forces F_{ri}, F_{ei}, $i = 1,2$, should be expressed in the form (5.26), (5.27).

Let us consider deformation. On both sides, there appear deformations along the horizontal and vertical directions. On the robot side, the external deformation has a horizontal and a vertical component: $s_r^{c1} = (s_{r1}^{c1}, s_{r2}^{c1})$. The internal deformation has the components: $s_r^{c2} = (s_{r1}^{c2}, s_{r2}^{c2})$. On the environment side, the horizontal and the vertical components of external deformation are: $s_e^{c1} = (s_{e1}^{c1}, s_{e2}^{c1})$. The internal deformation components are: $s_e^{c2} = (s_{e1}^{c2}, s_{e2}^{c2})$. Relation (5.34), connecting deformation and q-coordinates, now becomes:

$$\ddot{s}_{r1}^{c2} + \ddot{s}_{r1}^{c1} + \ddot{s}_{e1}^{c1} + \ddot{s}_{e1}^{c2} = \ddot{q}_e + J_{r11}^c \ddot{q}_{r1} + J_{r12}^c \ddot{q}_{r2} + A_1^c$$
$$\ddot{s}_{r2}^{c2} + \ddot{s}_{r2}^{c1} + \ddot{s}_{e2}^{c1} + \ddot{s}_{e2}^{c2} = J_{r21}^c \ddot{q}_{r1} + J_{r22}^c \ddot{q}_{r2} + A_2^c \ . \tag{5.47}$$

In the presented example we neglect the inertia associated with external deformation $(m_r^{(e)1} = m_e^{(e)1} = 0)$. This reduces the complexity of the elastodynamic model. The model of deformation on the environment side (equations (5.30), (5.26)) becomes

$$
\begin{aligned}
&F_i = -k_{ei}^1 s_{ei}^{c1} - b_{ei}^1 \dot{s}_{ei}^{c1} \\
&k_{ei}^1 s_{ei}^{c1} + b_{ei}^1 \dot{s}_{ei}^{c1} = m_{ei}^{(e)2} \ddot{s}_{ei}^{c2} + k_{ei}^2 s_{ei}^{c2} + b_{ei}^2 \dot{s}_{ei}^{c2} \\
&F_{ei} = -k_{ei}^2 s_{ei}^{c2} - b_{ei}^2 \dot{s}_{ei}^{c2} \\
&i = 1, 2 \ ,
\end{aligned}
\tag{5.48}
$$

and on the robot side (equations (5.31), (5.27)):

$$
\begin{aligned}
&F_i = -k_{ri}^1 s_{ri}^{c1} - b_{ri}^1 \dot{s}_{ri}^{c1} \\
&k_{ri}^1 s_{ri}^{c1} + b_{ri}^1 \dot{s}_{ri}^{c1} = -m_{ri}^{(e)2} \ddot{s}_{ri}^{c2} + k_{ri}^2 s_{ri}^{c2} + b_{ri}^2 \dot{s}_{ri}^{c2} \\
&F_{ri} = -k_{ri}^2 s_{ri}^{c2} - b_{ri}^2 \dot{s}_{ri}^{c2} \\
&i = 1, 2 \ .
\end{aligned}
\tag{5.49}
$$

In the models, i=1 indicates the horizontal direction, and i=2 the vertical direction. The models involve the contact force F (its horizontal component F_1 and vertical component F_2), which is essential in every contact task.

In this way we have completed the set of equations needed to describe the system dynamics and, thus, enable simulation.

Simulation

The robot parameters are given in Table 5.1. The links lengths are denoted by $l_j, j = 1,2$, masses by m_{rj}, and inertia moments by J_{rj}. The robot is driven by DC motors for which R_j is the armature resistance, I_j is the rotor inertia moment, C_{Mj} and C_{Ej} are the constants of torque and back e.m.f., and N_j is the gear ratio. It is assumed that the robot has static compensation that reduces gravity load by 70%, the remaining 30% being uncompensated. The cart mass is $m_e = 100\,kg$ and the motor is equal to those used for the robot drive. The motor shaft is directly connected to the cart wheel of the radius $r = 0.1\,m$.

The required velocity of the cart is $v_0 = 0.25\,m/s$. The robot starts from the resting position defined by $q_{r1}(0) = 0.$, $q_{r2}(0) = 5\pi/6$ (position A in Fig. 5.5). Then, the robot moves towards the cart. In the horizontal direction, the robot tip accelerates and achieves v_0 in 0.5 s. Then its horizontal velocity remains constant. In the vertical direction, the tip moves down with a triangular velocity profile so as to reach the level of the cart in 1 *s*. At this instant the contact with the cart is made (state B in Fig. 5.5). The motion is programmed so as to avoid impact (relative velocity should be zero). The robot has to produce the forces upon the cart: $F_1 = 100\,N$ in the horizontal and $F_2 = 100\,N$ in the vertical direction (state C in Fig. 5.5). It is planned that the forces attain the required value (100 *N*) exponentially.

Table 5.1. Robot parameters

	Links			Motors			Gears
j	l_j	m_{rj}	J_{rj}	R_j	I_j	$C_{Mj}=C_{Ej}$	N_j
1	1.	60.	5.	1.3	0.00079	0.5	200
2	1.	40.	3.5	1.3	0.00079	0.5	200

In the approaching phase (motion A→B), PD regulators were used to control the system. Dynamics was simulated by using the rigid-body model. The results of this phase are not so important for the topic and therefore are not presented.

After the instant of contact, elastodynamics is introduced into the model. Since the external deformation inertia is neglected, the impact shock does not occur. Parameters of the elastodynamic model are:

$$m_r^{(e)2} = m_e^{(e)2} = 10\ kg\,,$$
$$k_{ri}^1 = k_{ri}^2 = k_{ei}^1 = k_{ei}^2 = 10^4\ N/m\,,\ \text{and}$$
$$b_{ri}^1 = b_{ri}^2 = b_{ei}^1 = b_{ei}^2 = 10\ N/(m/s)\,.$$

In the contact phase of motion, dynamic position control was applied. Figs. 5.6–5.9 present the results of simulation. Fig. 5.6 shows the control voltages. The deformation in the horizontal direction is shown in Fig. 5.7, and the deformation in the vertical direction in Fig. 5.8. One can see that the external deformations converge to constant values, while the internal deformations are of the oscillatory nature. Such character of external deformations is due to the action of the control system (smooth external deformation directly implies smooth contact force). Fig. 5.9 shows the contact forces (in the horizontal direction and in the vertical direction).

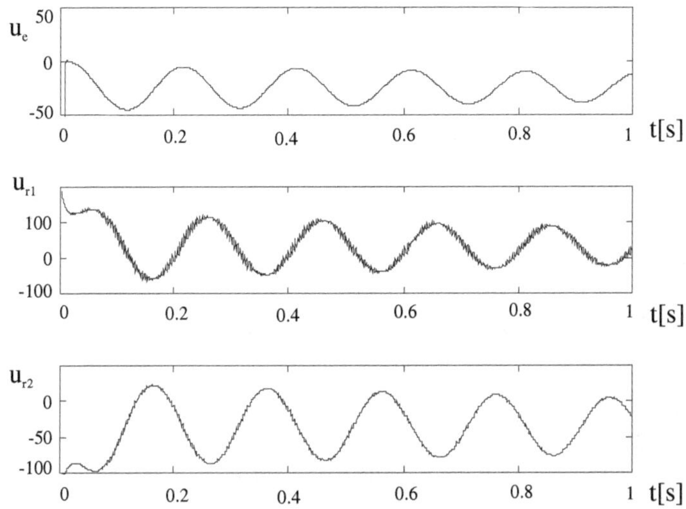

u_e - voltage for the cart motor
u_{r1}, u_{r2} - voltages for the robot motors

Figure 5.6. Control inputs.

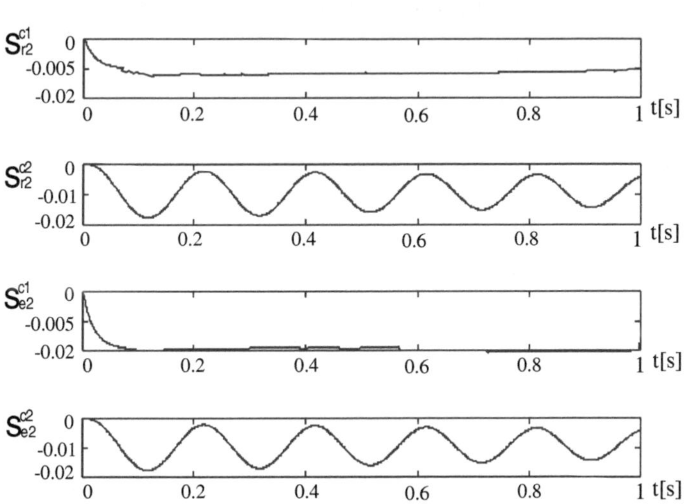

S_{r2}^{c1} - external deformation at robot side
S_{r2}^{c2} - internal deformation at robot side
S_{e2}^{c1} - external deformation at cart side
S_{e2}^{c2} - internal deformation at cart side

Figure 5.7. Deformation in the horizontal direction.

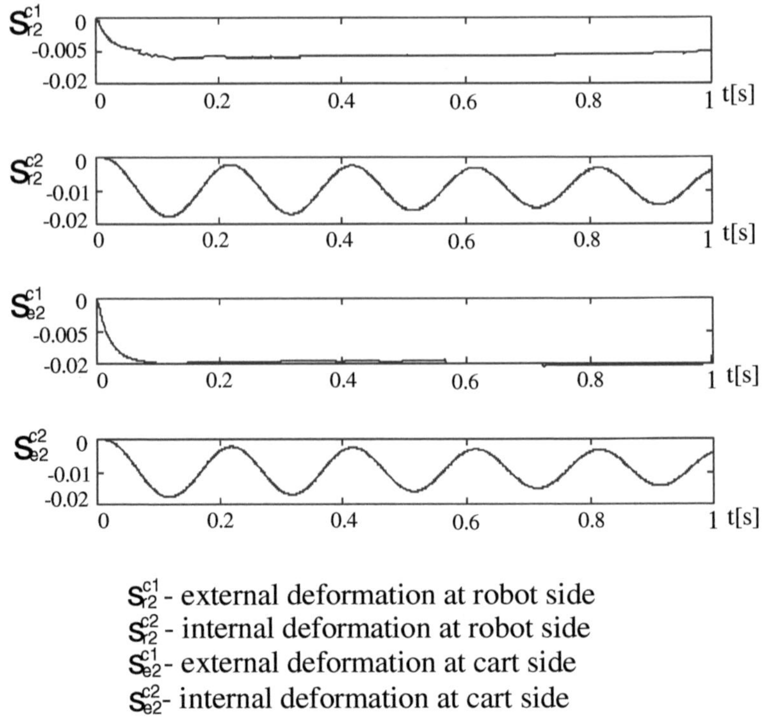

Figure 5.8. Deformation in the vertical direction.

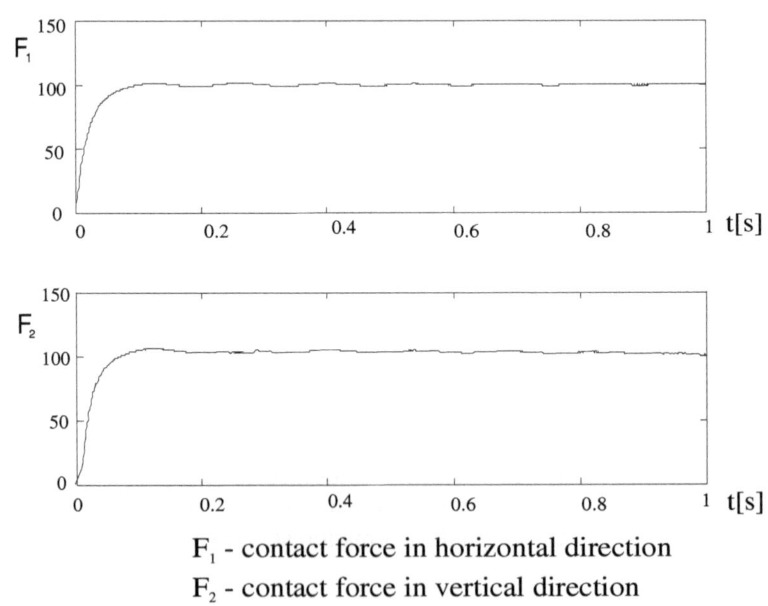

Figure 5.9. Contact forces.

4.2 Peg-in-Hole Assembly Task

We consider again an in-process operation. A robot has to insert a peg into a hole. The object with the hole is placed on a transport cart. Scheme of the task is given in Fig. 5.10. In this scheme, the peg is shown as being a part of the robot terminal link. The cart and the object with the hole are considered as one body and simply called the cart.

The robot has three DOFs. The data regarding its mechanism (geometry and inertial properties) are given in Table 5.2. Each joint is actuated by a DC motor.

The motors' parameters are: $C_M = 0.0575\,N/A$, $C_E = 0.0575\,V/(rad/s)$, $B_C = 1.432 \cdot 10^{-4}\,Nm/(rad/s)$, $R = 0.175\,\Omega$ and $I = 7.9 \cdot 10^{-4}\,kgm^2$, reduction ratio being $N = 50$.

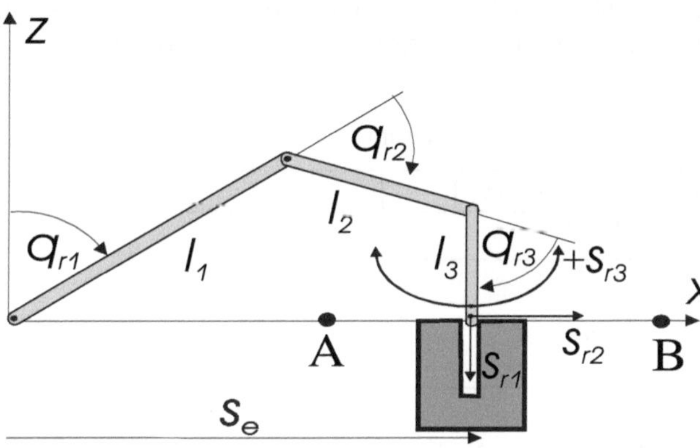

Figure 5.10. Peg-in-hole task. The task is being solved by a three-joint robot.

The cart represents the environment. Its mass is $m_e = 5\,kg$. It is driven by a DC motor that is equal to those used for the robot joint.

Let us explain the task. The cart moves horizontally from the position A $(x_A = 1m)$ to the position B $(x_B = 1.5m)$. The motion lasts $0.5\,s$, and is performed according to a trapezoidal velocity profile (acceleration and deceleration both take 20% of the motion time). The robot follows the motion of the cart and at the same time inserts its terminal link (the peg) into the hole. Insertion of $0.1\,m$ is performed in $0.5\,s$ with the trapezoidal velocity profile (acceleration/deceleration time is 20%). During insertion, friction occurs between the peg and the hole. The friction coefficient is $\mu = 0.4$. It is clear that the insertion requires vertical direction of the peg (terminal link). Besides, it is desired to perform insertion with the contact force equal to zero.

Now, we define the position coordinates. For the cart, the functional *s*-coordinate coincides with the *q*-coordinate and both coincide with *x*:

$s_e = q_e = x$. For the robot, the joint coordinates are q_{r1}, q_{r2}, and q_{r3}. Functional coordinates are selected so as to define the relative position with respect to the cart. The origin of the frame is placed at the brim of the hole. Two coordinates will define the position of the tip with respect to the origin, and the third coordinate will define the angular orientation of the terminal link. The notation is as follows: s_{r1} represents the insertion, s_{r2} defines the horizontal distance of the tip (measured from the hole brim), and finally, s_{r3} represents the angular deviation of the terminal link from the vertical direction oriented downwardly. The complete vector s includes the one-component vector s_e and the three-component vector s_r: $s = (s_e, s_{r1}, s_{r2}, s_{r3})$. Now, one should make a new partition of the vector s into free and constrained coordinates. Looking at the task illustrated in Fig. 5.10, one can find that the cart motion and the peg insertion represent the free coordinates: $s^f = (s_e, s_{r1})$. The contact between the peg and the hole constrains the lateral motion and the angular motion of the peg, and thus: $s^c = (s_{r2}, s_{r3})$.

Table 5.2. Parameters of the robot mechanism

J	$l_j\,[m]$	$m_{rj}[kg]$	J_{rj}
1	1.0	20	$J_{rj} = \frac{1}{12} m_{rj} l_j^2$
2	0.6	10	
3	0.4	5	

We shall discuss two cases, the rigid-body contact and the elastodynamic contact. In any case, the cart is not engaged in controlling the contact. It is forced to track the reference motion by applying the PID regulator. For the robot, two kinds of control will be examined: classical hybrid position/force control and dynamic position control (see Appendix 1 and Appendix 2).

The rigid-body contact

The rigid-body contact is discussed first. It is expressed by the requirement: $s^c = 0$, that is, $s_{r2} = 0$ and $s_{r3} = 0$. For controlling the robot, the classical hybrid law is applied. The theory explained in Sections 5.1 and 5.2 allows simulation of the system behavior. Strictly speaking, there must exist an approaching phase, then the impact phase, and following that the "regular" insertion starts. Since the approaching is not of special interest here, and the impact has already been discussed in detail in several examples, we start presentation of the results from the moment when the impact ended. At that instant the peg tip is inserted into the hole for 3 *mm*. So, $s_{r1}(0) = 0.003$ represents the initial position. However, the reference motion assumes the insertion of 5 *mm* in that instant. This means

$s_{r1}^{*}(0) = 0.005$ and, thus, the initial error is $\Delta s_{r1}(0) = s_{r1}^{*} - s_{r1} = 0.002$. The mark * indicates the reference values. This error follows from the tracking error in the approaching phase as well as from impact effects.

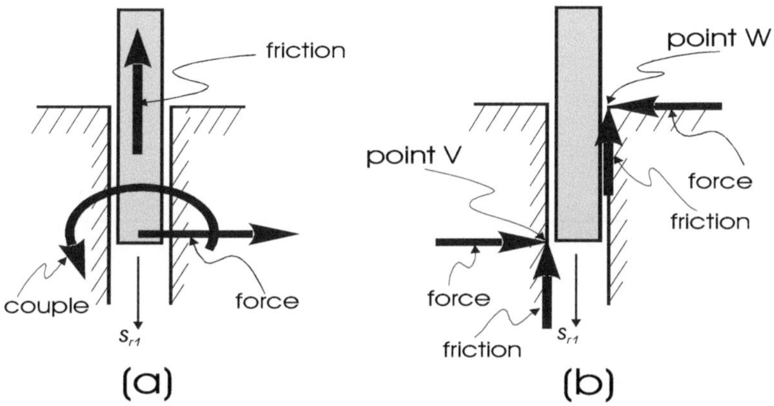

Figure 5.11. Two approaches to the contact action/reaction: (a) one force and one couple express the interaction of the robot and the cart, (b) two forces are used to express the interaction.

Simulation encompasses the calculation of the cart and robot motion as well as the calculation of contact force. The motion is described either in terms of the internal q-coordinates: q_e for the cart and q_{r1}, q_{r2}, and q_{r3} for the robot, or the functional s-coordinates: s_e for the cart and s_{r1} for the robot (note that $s_{r2} = 0$ and $s_{r3} = 0$). The analysis of forces and their presentation requires some additional explanation. Contact force appears along the direction of each constrained coordinate. Hence, the contact action (or reaction) has two components: a force along the lateral direction s_{r2} and a couple along the angular direction s_{r3} (see Fig. 5.11a). When the peg-in-hole task is considered, it may be more convenient and more similar to a real application if contact is considered in terms of the two forces shown in Fig. 5.11b. The number of unknowns does not change, as the two forces replace a force and a couple. This approach will be even more beneficial when we discuss the elastodynamic contact. Regardless of the representation of contact, the friction force is introduced, acting in the opposite direction to the free coordinate s_{r1} (Fig. 5.11 a, b).

Figs. 5.12–5.15 present the simulation results. Fig. 5.12 shows the motion of the cart. Time histories of the coordinate q_e, the tracking error Δq_e, and the velocity $\dot{q}_e$, are given. Fig. 5.13 shows the robot motion in terms of its internal coordinates q_{r1}, q_{r2}, and q_{r3}. Fig. 5.14 shows the insertion, giving time histories of the relative coordinate s_{r1} and the tracking error Δs_{r1}. Figure 5.15 shows the contact forces F_v and F_w at the points V and W, and the total friction force F_{fr}.

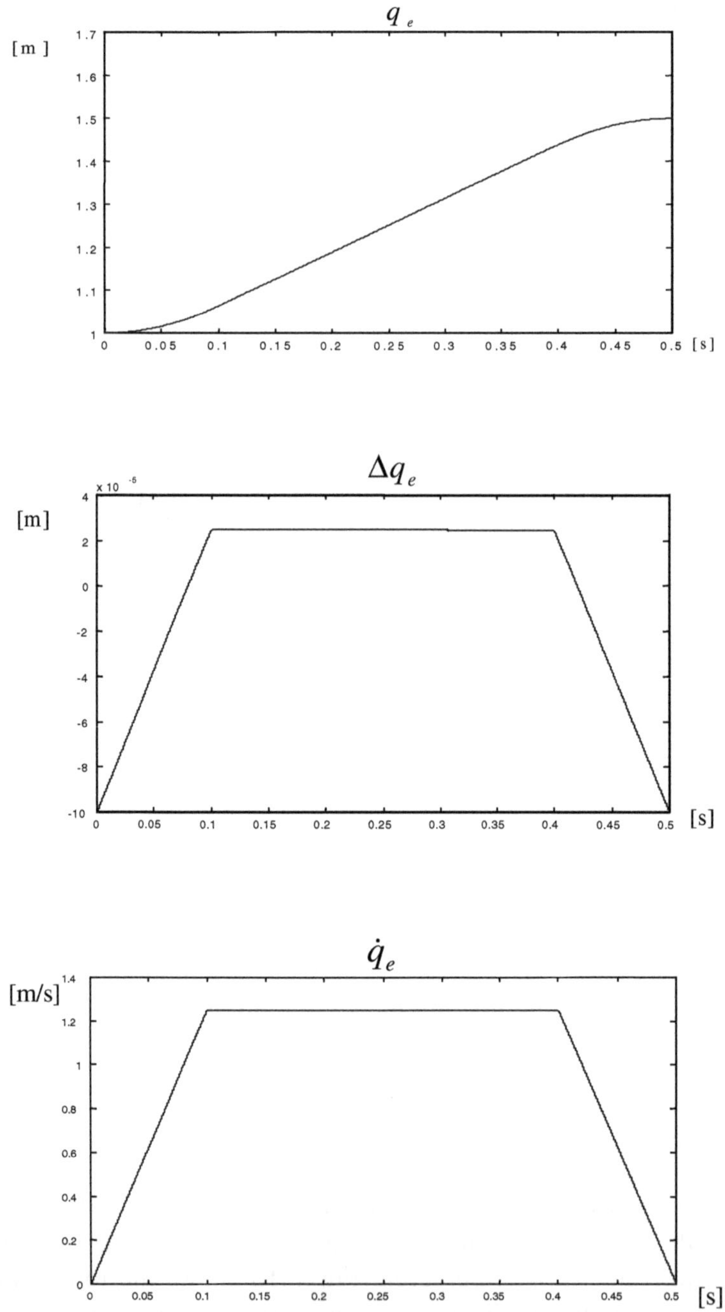

Figure 5.12. Simulation of rigid-body contact and classical hybrid control: Time histories of cart motion (q_e), tracking error (Δq_e), and velocity $(\dot{q}_e)$.

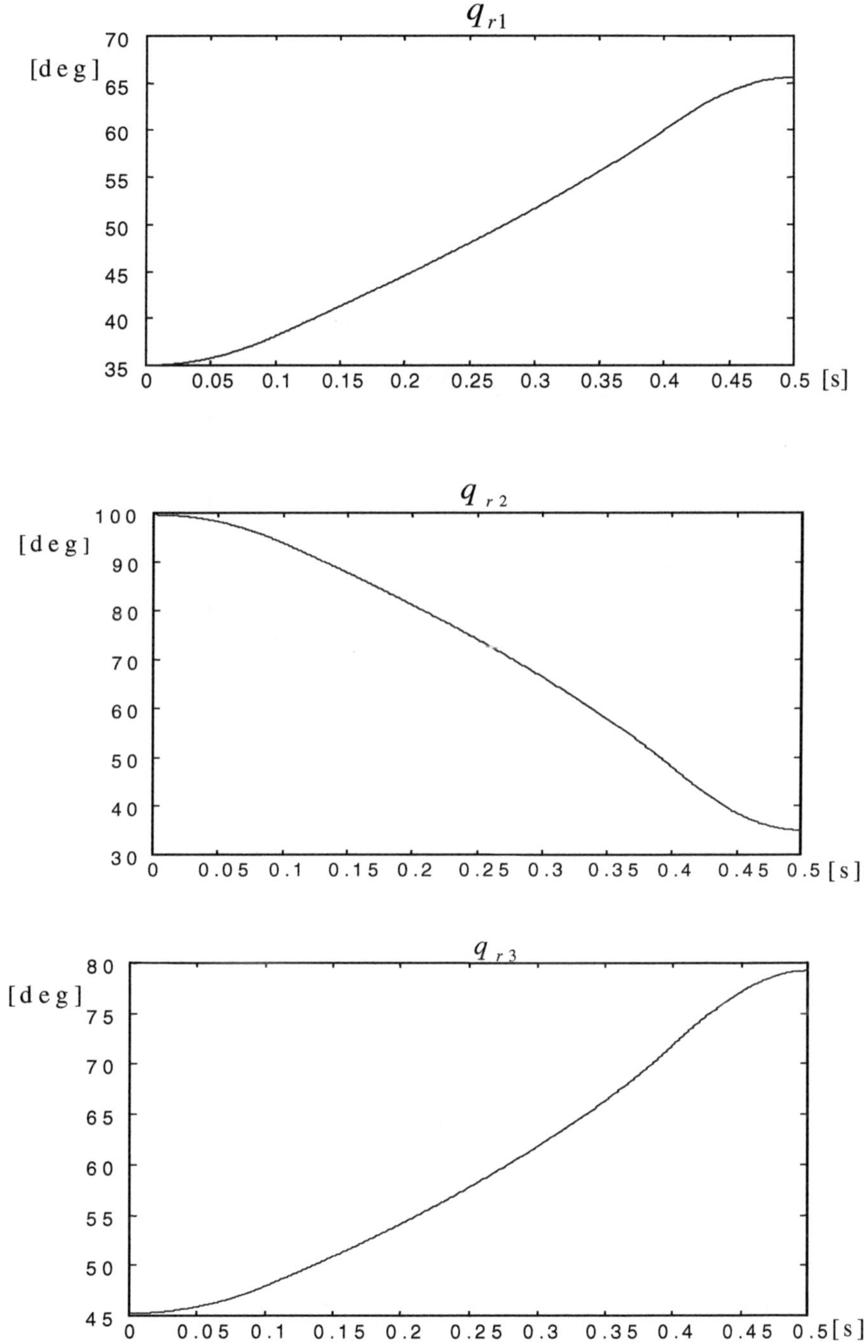

Figure 5.13. Simulation of rigid-body contact and classical hybrid control: Time histories of robot joint coordinates q_{r1} , q_{r2} , and q_{r3} .

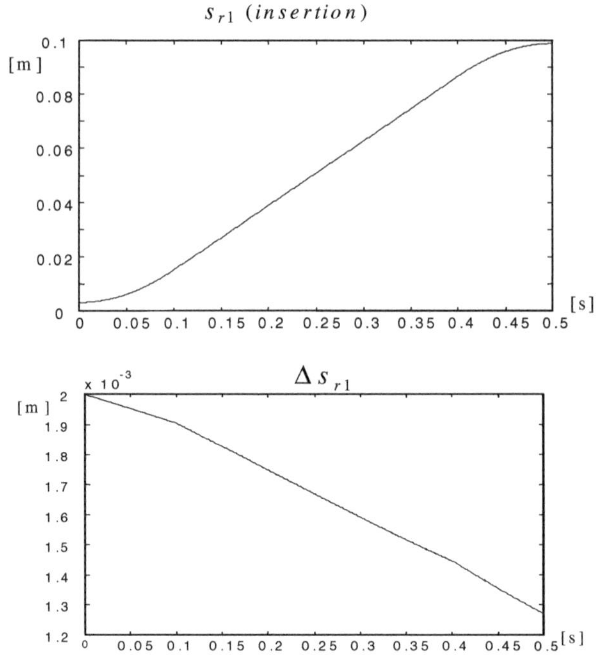

Figure 5.14. Simulation of rigid-body contact and classical hybrid control: insertion coordinate s_{r1} and its tracking error Δs_{r1}.

The elastodynamic contact

Now we introduce elasticity in the contact zone. It will be assumed that the peg is infinitely rigid, and thus, deformation appears on the cart side only. In the rigid-body consideration, the insertion, being a two-component contact, was described by zero values of the two coordinates s_{r2} and s_{r3}. When deformation is introduced, this does not hold. Theoretically, the two coordinates may take arbitrary values, but the strong elastic forces will keep these values small. Thus, lateral and angular deviations of the peg position, although small, should be taken into account. However, we decided to express the two-component contact in terms of the two contact points (points V and W in Fig. 5.11b).

In the case of contact with deformation, there appears an elastic deflection at each contact point. So, instead of the deviations s_{r2} and s_{r3}, we introduce the deflections v and w, as shown in Fig. 5.16. Each deflection is characterized by an equivalent mass $m^{(e)}$, stiffness k, and the damping coefficient b. This means that one-stage deformation model is used. For simulation we adopted: $m^{(e)} = 1kg$, $k = 10^5 N/m$, and $b = 1N/(m/s)$.

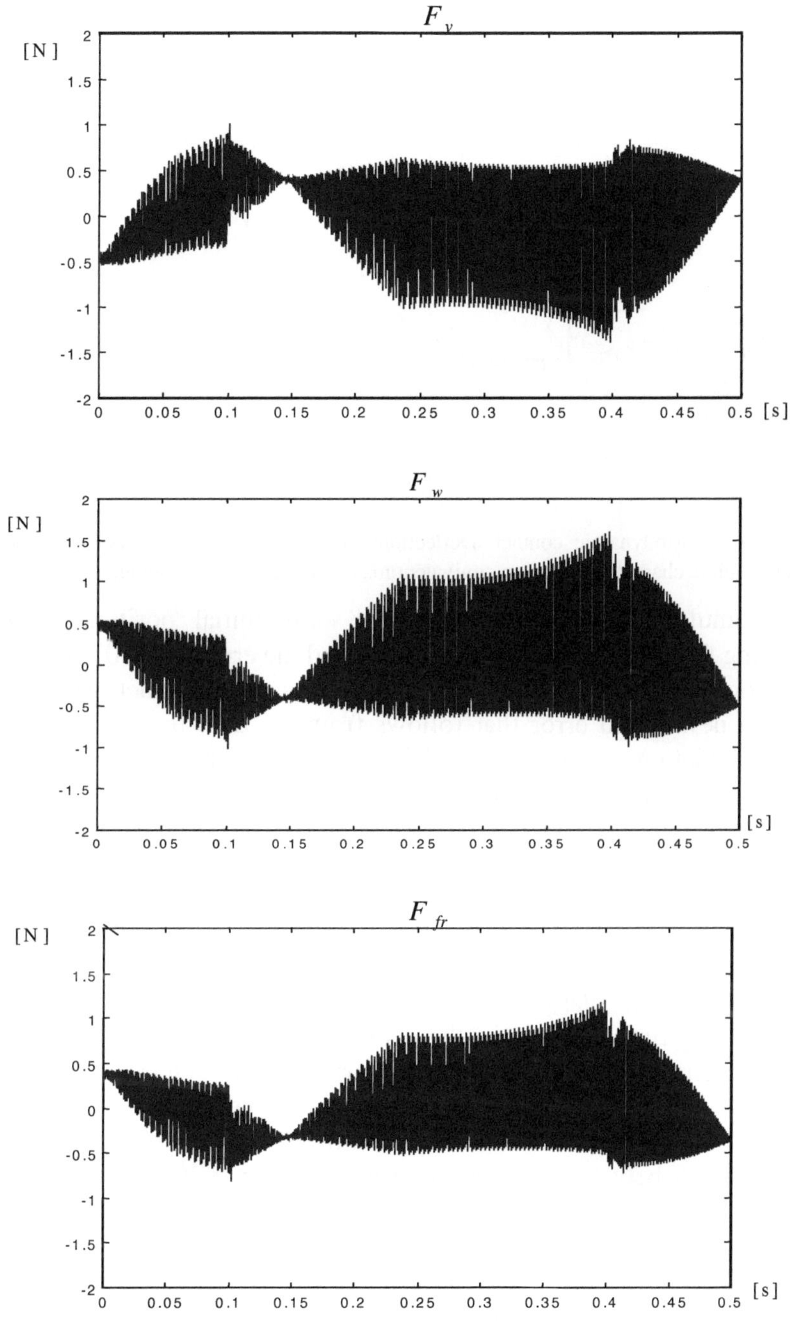

Figure 5.15. Simulation of rigid-body contact and classical hybrid control: contact forces F_v and F_w, and total friction force F_{fr}.

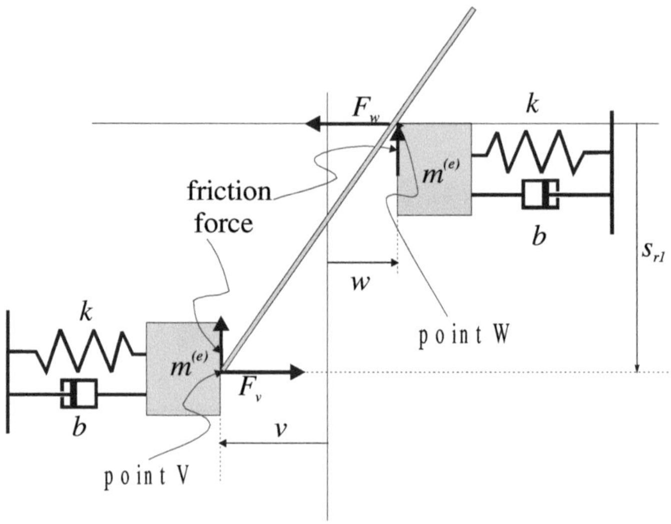

Figure 5.16. Elastodynamic contact. Deflections at the two contact points are introduced. Each deflection is characterized by an equivalent mass, stiffness, and damping.

The simulation started by adopting some initial position error. In considering the rigid-body contact we adopted the error $\Delta s_{r1}(0) = 0.002$ in the direction of insertion. In the case of elastodynamic contact, we keep this and add a new initial error that follows from the deformation of the hole: angular deviation $s_{r3}(0) = -3.81\deg$ follows from the deflections $v(0) = 0.1mm$ and $w(0) = 0.1mm$.

For controlling the robot in contact motion, two concepts were checked and compared. First, the classical hybrid position/force control law was applied, and second, the dynamic position control. Let us remember that the cart is not engaged in contolling the contact – it is forced to track the reference motion by applying the PID regulator.

We start with the classical hybrid control. Fig. 5.17 shows the motion of the cart, giving time histories of the coordinate q_e, the tracking error Δq_e, and the velocity $\dot{q}_e$. Fig. 5.18 shows the robot motion in terms of its internal coordinates q_{r1}, q_{r2}, and q_{r3}. Fig. 5.19 displays the insertion, giving time histories of the relative coordinate s_{r1} and the tracking error Δs_{r1}. To allow the comparison between the two control laws, Fig. 5.20 presents insertion ($s_{r1}, \Delta s_{r1}$) in the case of dynamic position control. Figs. 5.21 and 5.22 show the deformation: 5.21 for the case of classical hybrid control and 5.22 for the dynamic position control. The deflections v and w are shown as well as the angular deviation of the peg (s_{r3}) which is the consequence of these two deviations. Figs. 5.23 and 5.24 show the contact forces F_v and F_w, and the total friction force F_{fr} (Fig. 5.23 holds for the classical hybrid control, and Fig. 5.24 for the dynamic position control).

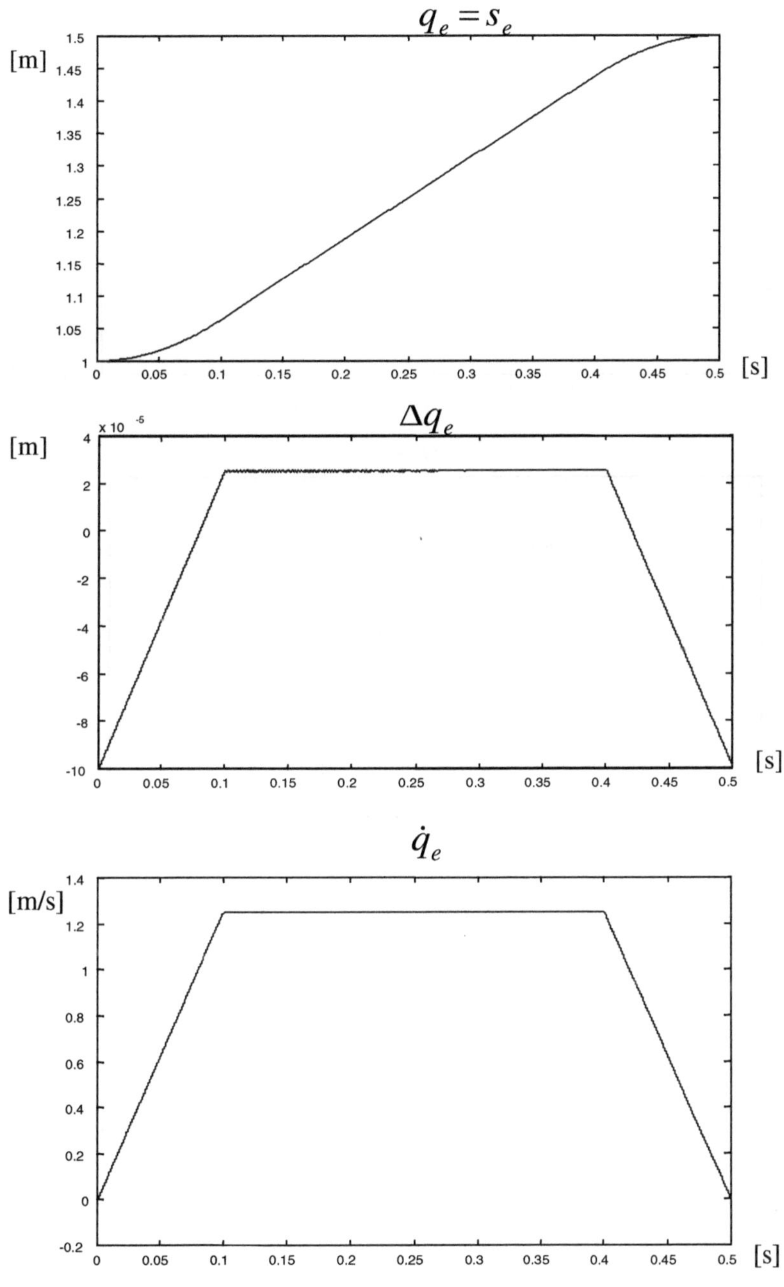

Figure 5.17. Simulation of elastodynamic contact and classical hybrid control: time histories of cart motion (q_e), its tracking error (Δq_e), and velocity $(\dot{q}_e)$.

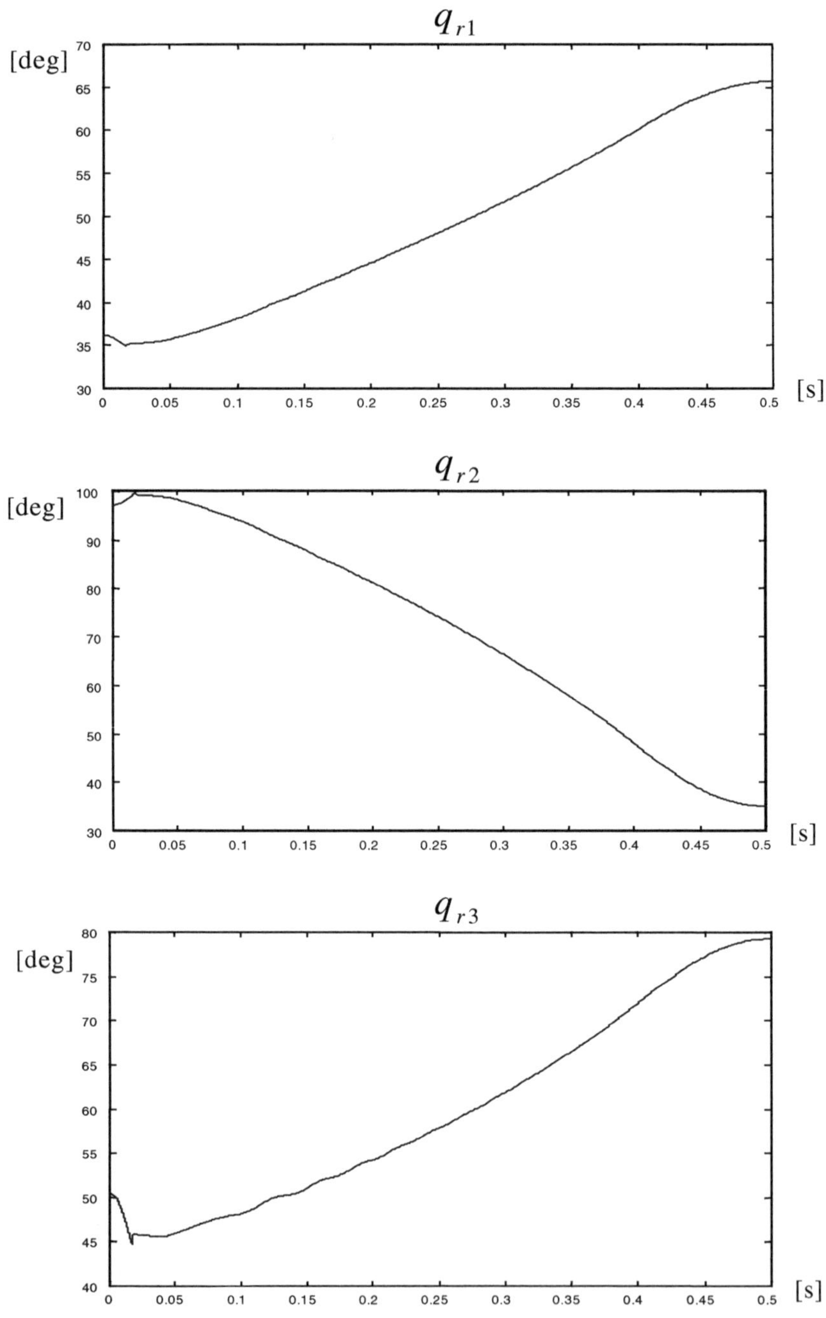

Figure 5.18. Simulation of elastodynamic contact and classical hybrid control: time histories of the robot joint coordinates q_{r1}, q_{r2}, and q_{r3}.

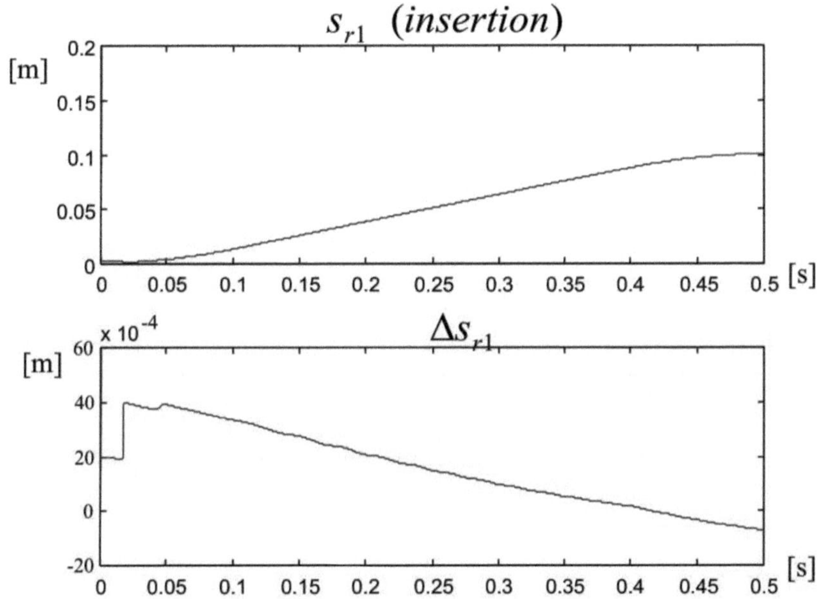

Figure 5.19. Simulation of elastodynamic contact and classical hybrid control: insertion coordinate s_{r1}, and its tracking error Δs_{r1}.

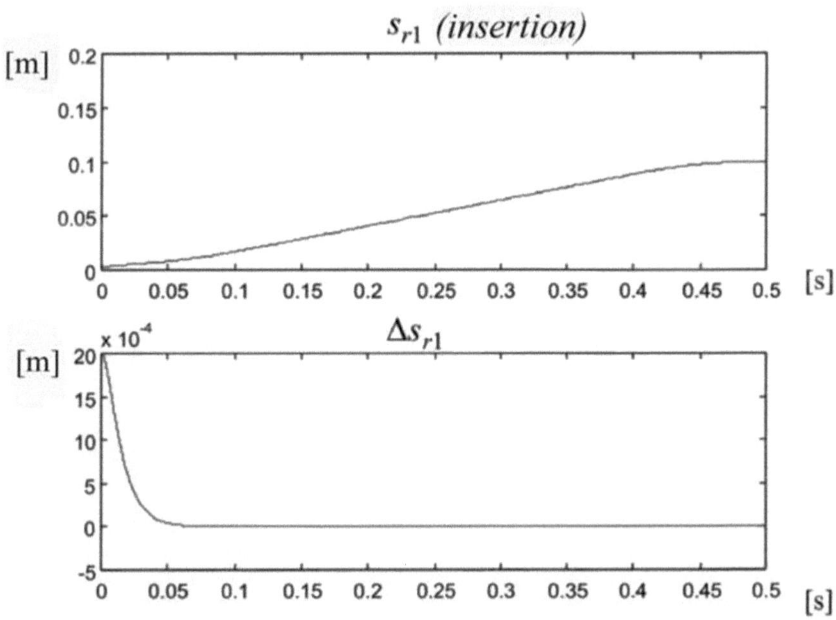

Figure 5.20. Simulation of elastodynamic contact and dynamic position control: insertion coordinate s_{r1}, and its tracking error Δs_{r1}.

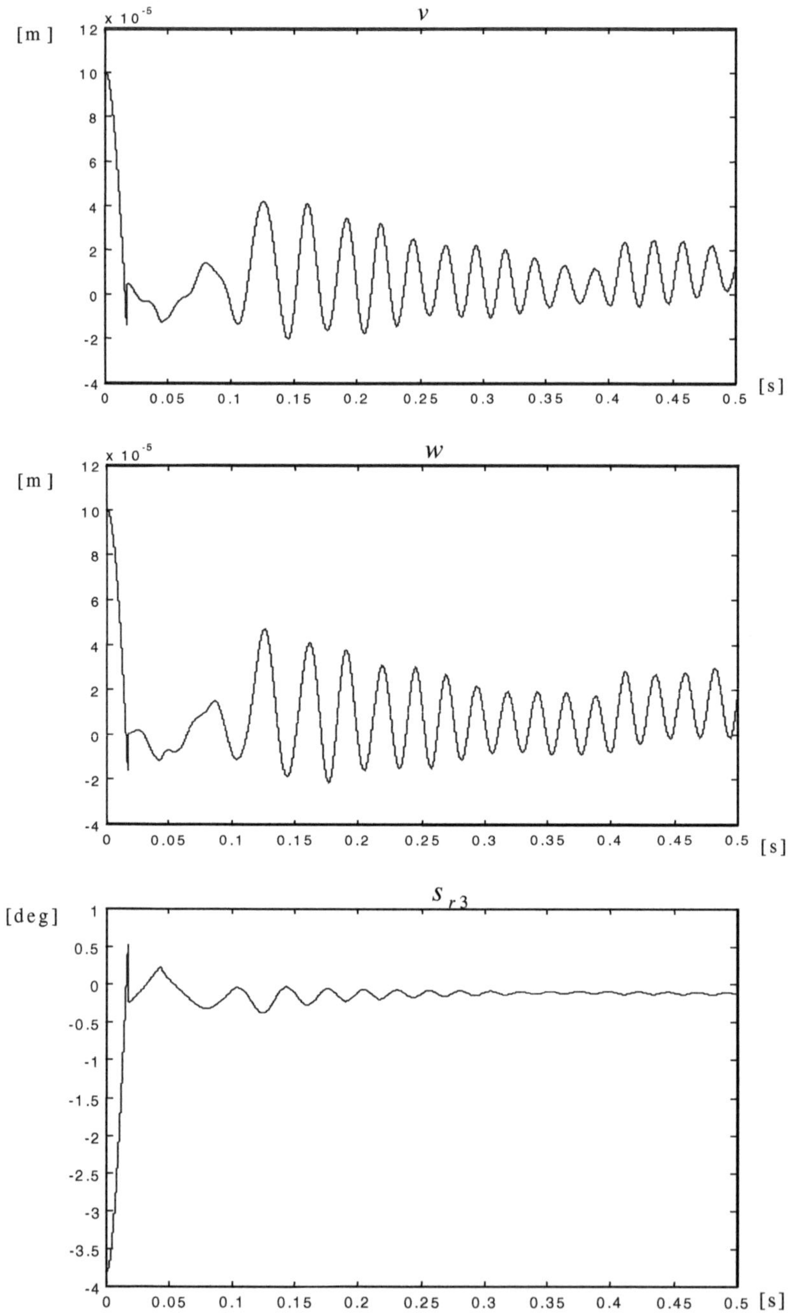

Figure 5.21. Simulation of elastodynamic contact and classical hybrid control: deflections at the contact points $-v$ and w, and angular deviation of the peg.

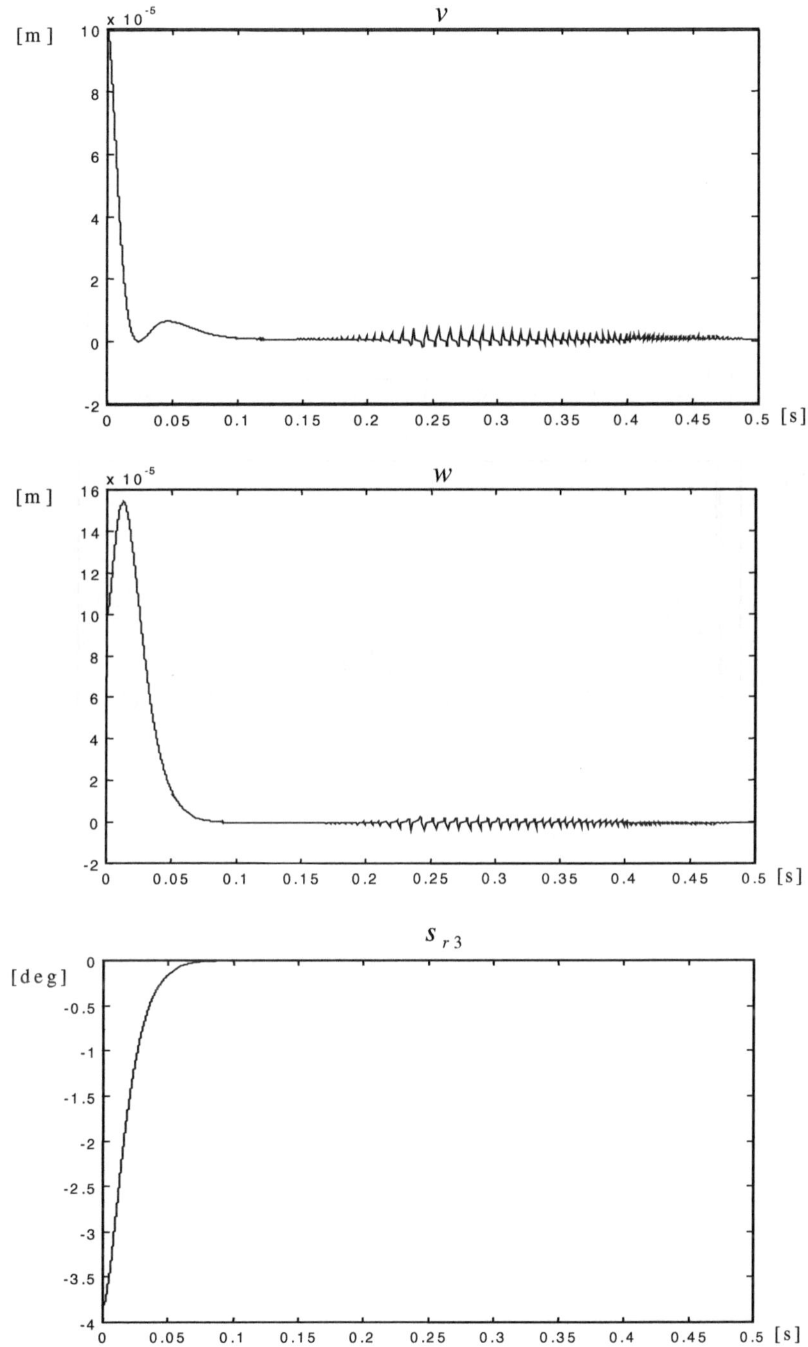

Figure 5.22. Simulation of elastodynamic contact and dynamic position control: deflections at contact points $-v$ and w, and angular deviation of the peg.

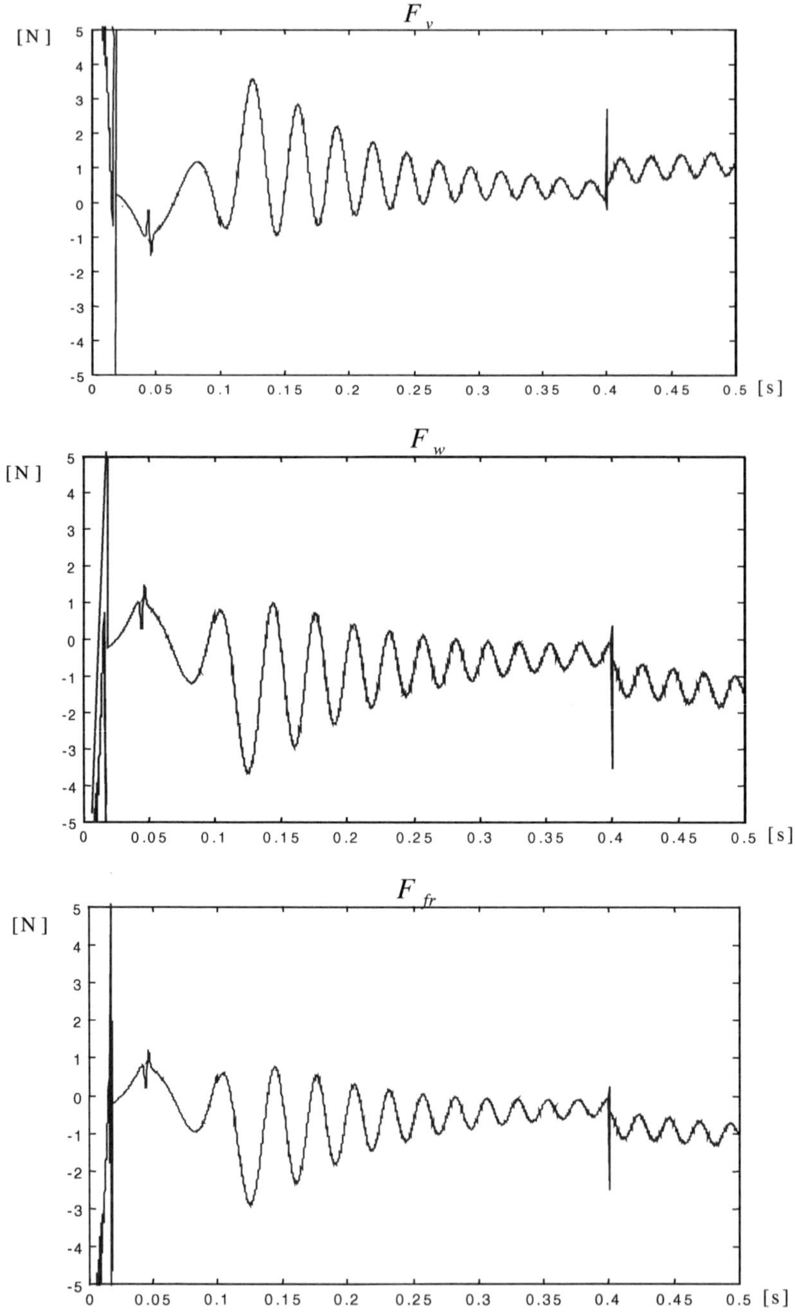

Figure 5.23. Simulation of elastodynamic contact and classical hybrid control: contact forces F_v and F_w, and total friction force F_{fr}.

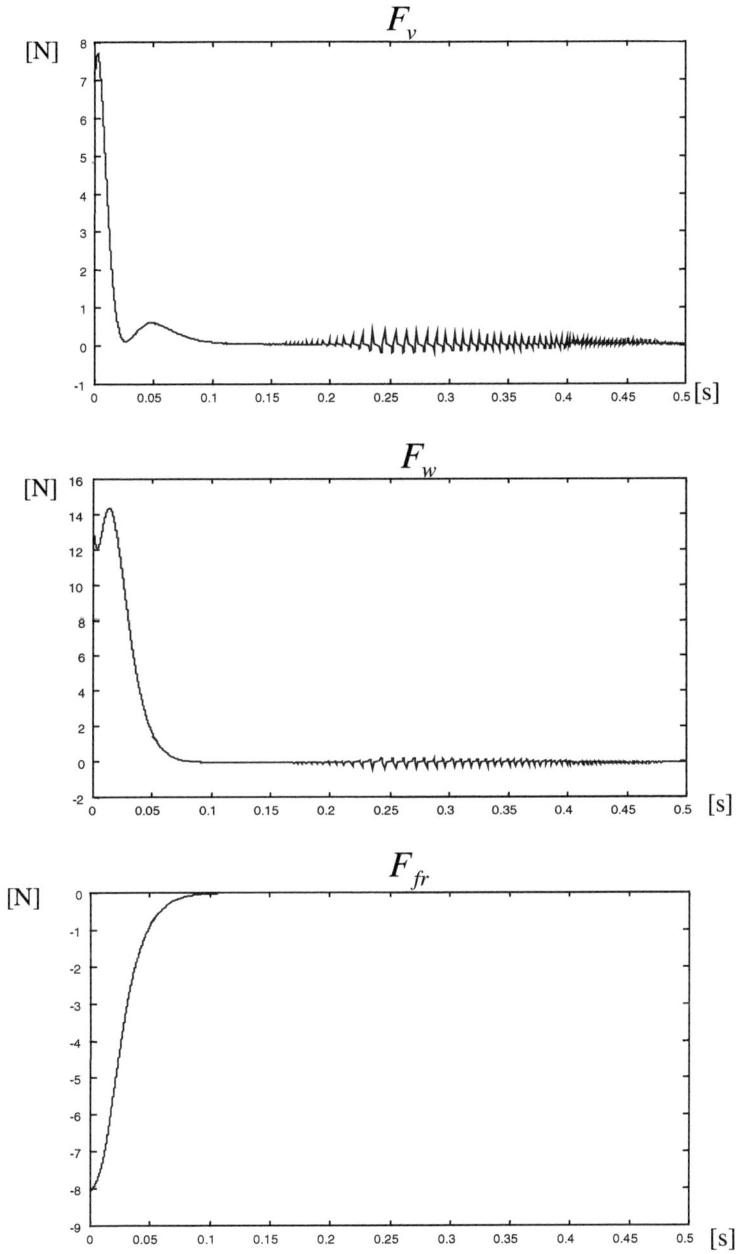

Figure 5.24. Simulation of elastodynamic contact and dynamic position control: contact forces F_v and F_w, and total friction force F_{fr}.

Chapter 6

DYNAMICS OF MULTI-ARM COOPERATIVE ROBOTS WITH ELASTIC CONTACTS

1. FUNDAMENTALS

In this Chapter, the procedure of modeling and the complete general form mathematical model of manipulators with six degrees of freedom (DOF) in cooperative work are presented, together with the solution of the indefiniteness problem with respect to force distribution. The obtained model is presented in several convenient forms. For the first time, a system of active spatial six-DOF mechanisms elastically interconnected with the object (dynamic environment) is modeled. The reason for the emergence of the indefiniteness problem with respect to force is explained and the procedure for solving this problem is given. Unlike the approaches given in the available literature, the indefiniteness problem with respect to force is solved in accordance with physical phenomena.

1.1 Problem Definition

An arbitrary number of six degree-of-freedom (DOFs) rigid manipulators, performing cooperative work on a single manipulated object is considered. The motion of the manipulated body is constrained only by the kinematic properties of the manipulators themselves. The elasticity of manipulators is much less than the elasticity of the object, therefore the manipulators can be assumed to be rigid. The elastic properties of the object are assumed to be either elastic in the contact area only (the rest of the object being rigid) or uniformly elastic over the whole object body. In both cases it is assumed that each connection or part of the manipulated object body in the vicinity of the contact point, can be represented as a local rigid body in the mass center (MC) of which contact, gravitational, damping, and elasticity forces act. Thus, the manipulated object body with m manipulator connections can be thought of as the system consisting of $m+1$ elastically interconnected rigid bodies (Fig. 6.1). It is considered that each locally rigid body has six DOFs. The gravitational and contact forces are considered as the external ones.

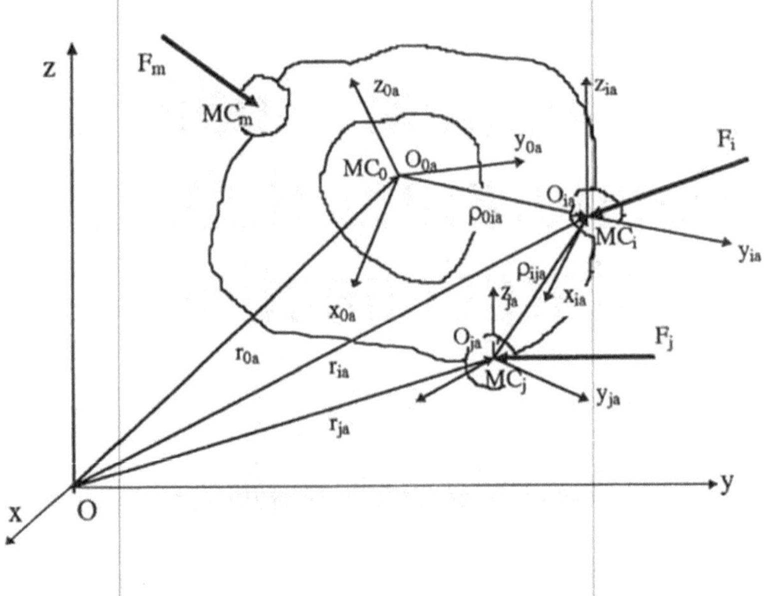

Figure 6.1. Elastic system of $m+1$ rigid bodies.

The problem is posed as follows: Knowing the mutual positions and attitudes of the manipulator tips at the moment when the contact with the manipulated object is made, derive the mathematical model of the above-defined dynamic system as a function of the absolute positions of the contact points and the manipulated objects MC.

1.2 Some Known Solutions of the Cooperative Manipulation Model

The basic problems in solving the cooperative work lay in determining the forces at the contact point of the manipulator tip and the object, and in determining the position of the object/manipulator based on the knowledge of the positions of the manipulator/object.

These problems can be defined as those related to the choice of the assumptions about system behavior, and those related to the quality of mathematical description of cooperative work obtained through the introduction of these assumptions. In the majority of papers related to cooperative work it is assumed that manipulators and objects are rigid. As a rule, a consequence of this assumption is a model with an undefined force, which reflects itself in the impossibility of finding a unique relation between the force vector at the MC of the manipulated object and the vector of forces at the contact point of the manipulator and object. In trying to solve this problem different approaches have been made, but it is common to all of them that the forces at the contact point are determined by the conditions which are not the consequences of the physical phenomena, but of the assumptions made by the authors [1–7]. These conditions were given from the standpoint of demands concerning the object, the manipulator, or their combinations.

In [1] and [2], a vector f_c at contact points was chosen completely independently of the manipulator dynamics, i.e. the criterion for the choice of vector f_c was based on the requirements concerning the object. The solution for the forces at contact point was chosen so that it minimizes the quadratic criterion

$$I_f = f_c^T W f_c \tag{6.1}$$

and the solution is obtained in the form

$$\begin{aligned} f_c &= W^{-1}H^T(HW^{-1}H^T)^{-1}F_o, \\ f_{ci} &= W_i^{-1}\left(\sum_{j=1}^{n} W_j^{-1}\right)^{-1}F_o, \quad i=1,\dots,n \end{aligned} \tag{6.2}$$

where W=diag (w_1, ..., w_n) is a weighting matrix, F_O is a force vector in the object MC, and $H=(I\ I \dots I)$ is a block of unit matrices resulting from the relation $F_o = \sum f_c = Hf_c$. A possible practical solution for the redistribution of the load f_c for "slave" manipulators was given in the form of the internal forces vector f_I [3]. In [2], the distribution was adopted as a solution, by minimizing the functional min $\{\|f_c\|\}$ along with satisfying the statical friction conditions expressed by the inequality $e_{Ni} f_{ci} \geq \eta_i \|f_{ci}\|$, where $e_{Ni} = grad\, S(p_i) / \|grad\, S(p_i)\|$ is the vector of the normal at point p_i, onto the object surface described by $S(x,y,z)=0$, and η is the friction coefficient. In [4], requirements for the internal forces were chosen in such a way so as to preserve the contact force inside the friction cone.

The solution of redundancy was also sought independently of the object dynamics, ([8], [5]). In [8] the functional $\tau W \tau$ was minimized and the solution, which is by its form analogous to the solution of the contact forces functional, was obtained. The possible optimal distribution of load between two industrial robots was considered in [5].

As one possible solution for drives, τ was proposed, satisfying the condition $OBJ > \|\tau\|^2$, where OBJ represents the minimum energy criterion. For uniform load distribution it was proposed that the contact forces be equal and also equal to a half of the force at the object MC.

The approach where both the object and manipulator dynamics were taken into consideration is also possible. In the case of absence of constraints of the "leader" and "follower" driving torques, as in [5], the solution equivalent to minimum energy clause is obtained by combining the right-hand sides of the equations for object and manipulator behavior, and by minimizing the driving vector norm. The result is a practically inapplicable solution for the driving torques in the form $\tau^{l/f} = \tau^{l/f}(\tau, F_0, f_c)$, due to the copious computations needed. The distribution $f_c^l = \alpha F_0$, $f_c^f = (1-\alpha) F_0$, $0 < \alpha < 1$ was considered as an alternative procedure. The contact forces vector norm was minimized $\min\left\{ \|f_c\| \right\} = \min\left\{ \left(\|f_c^l\|^2 + \|f_c^f\|^2 \right) \right\}$ and the solution was obtained as the function of internal forces f_I as $\alpha_r = \alpha_r(\tau, F_0, f_I)$.

In [9], the driving torques were presented in the form $\tau = \tau' - J^T(I - G^+ G)\varepsilon$, where τ' represents the drive ensuring motion along the trajectory, G represents the matrix of the expanded vector transform action of the contact points velocity into the velocity vector of the object MC, G^+ represents the generalized pseudo-inverse matrix (Moore–Penrose) of matrix G, and ε represents an arbitrary vector. Vector ε was chosen whereby there was given the possibility of supervising the internal forces. The following equation was stated as a possible choice: $\varepsilon = \mathrm{sgn}(\tau')\left[J^T\right]_i^T$, where $\left[J^T\right]_i^T$ is the $i-$th row of transformation matrix J of the internal coordinates velocity vector into the expanded vector of contact points velocity, which leads to the manipulator load reduction. In the case when cooperative system mobility is higher than the dimensions of the object operative space, vector ε was chosen by satisfying condition $[P]$ of some subtask described in form $[P]\tau = \alpha$. As a solution we obtain: $\varepsilon = \left([P] - J^T(I - G^+G)\right)^+ (\alpha - [P]\tau')$, representing the generalization of the approach given in [5]. As formally regarded, as long as the system has not been definite in a mathematical sense, new equations are added to the differential equations of cooperative manipulation.

1.3 Fundamentals of the Proposed Methodology for Cooperative Manipulation Modeling

The assumption that the manipulators and object are rigid, results in the condition of nonelastic contact between them and force indefiniteness. The unique solution of cooperative work model requires, then, the introduction of additional assumptions. In other words, supplementary natural conditions are needed in which the model will be completely mathematically defined.

For the system at rest, the driving torques and the forces at the point of contact of the manipulator tip and object can be considered as a system of internal generalized forces, while the gravitational forces can be considered as a system of external forces. Then, the cooperative system corresponds to a statically undefined grid with sticks clamped in the supports, the other ends of which are connected by an object. The ways of solving such grids have been elaborated in detail in [10–13]. For conditions at rest, results obtained in cooperative work model should be in accordance with the results already obtained in other branches of mechanics (statics, dynamics, materials resistance, and constructions theory). However, the results are in accordance with statics only, and due to the assumption of rigid manipulators and object, force indefiniteness appears.

Force indefiniteness can be overcome if the assumption of rigid manipulator and object is abandoned, or if that assumption is kept, but an elastic connection is established between the rigid manipulator and rigid object, along with deformations compatibility condition fulfilment. According to that condition, each construction is deformed in such a way that even after this deformation the connections between the individual construction elements are preserved. As many times as the construction is undefined, supplementary explicit geometric conditions appear, from which for known forces the unknown displacements are derived, and vice versa. As a result, unique values of forces/moments and displacements of construction are obtained in all its points (cross sections). In other words, by omitting the rigid body assumption, the necessity of introducing the vague additional conditions and geometrical restraints is avoided.

Several methods exist for solving statically undefined problems. The deformation work method will be considered, which originated from the principle of minimum potential system energy. In the science of materials resistance it is supposed that during deformation neither the change of heat energy nor acceleration of any material particle occur, i.e. it is supposed that the load changes are very slow, therefore according to the energy conservation principle for any elastic system, the equation $\delta A_d + \delta Q = \delta U + \delta T$

reduces to the equality of the deformation work and potential energy increments $\delta A_d = \delta U$, where δA_d is the increment of external forces work, δQ is the heat quantity increment, δT is the kinetic energy increment, and δU is the internal potential energy increment (i.e. deformation work increment).

For small displacements of the elastic system, or for displacements in the region of linear function of stress and deformation, the deformation work has a homogeneous quadratic form of either the forces (external and the 'statically unknown') or displacements u.

$$A_d = \frac{1}{2}\sum_{i=1}^{n}\sum_{j=1}^{n}\alpha_{ij}F_iF_j = F^TWF \tag{6.3}$$

$$A_d = \frac{1}{2}\sum_{i=1}^{n}\sum_{j=1}^{n}\beta_{ij}u_iu_j = u^TKu \tag{6.4}$$

where: W is the matrix of the so-called "Maxwell's displacement influence numbers" α_{ij}, which represent the displacements of the force F_i point of attack onto the direction of that force due to the unit force F_j, K is the matrix of stiffness or the matrix of "Maxwell's dual (reciprocal) coefficients" β_{ij}, representing the force which, acting at point j, produces unit displacement at point i, whereby the displacements at other points are equal to zero, and u_i are the corresponding displacements.

By means of the first and second statement of Castigliano, the displacements and forces are determined as derivatives of deformation work with respect to forces and displacements, respectively

$$u_i = \sum_{k=1}^{n}\alpha_{ik}F_k = \frac{\partial A_d}{\partial F_i} = W_i F \tag{6.5}$$

$$F_i = \sum_{k=1}^{n}\beta_{ik}u_k = \frac{\partial A_d}{\partial u_i} = K_i u \tag{6.6}$$

where W_i and K_i are the $i-$th rows of matrices W and K, respectively.

Let it be noted that the deformation work and deflexions are inversely proportional to the elasticity module. Comparing (6.3) and (6.4) with the attempts to resolve redundancy in cooperative work, it can be said that criterion (6.1) given in [1] alone looks most like the expression for deformation work, but it does not present the deformation work. It presents, rather, an arbitrarily chosen criterion with weighting elements matrix, and not the "Maxwell's displacement influence numbers". Even in the case when this criterion would represent deformation work, it could be correctly applied only for static conditions of cooperative work, but in that case the forces at the contact points of the manipulator tips and the object, as internal forces of the grid, cannot be determined according to (6.2) [1], but only according to (6.6).

Based on (6.5) and (6.6), several important conclusions impose themselves.

There exists a unique functional relation between displacement and force. This relation is linear for small displacements and for displacements in the region of linear dependence of stress and dilatation. If it is desired to realize a force F_i, given in advance, then the corresponding displacements (deflexions), i.e. the grid position (of the cooperative system) with respect to the unloaded system position should be realized, i.e. with respect to the position of establishing the contact in which all contact forces of the manipulator tip and object are equal to zero. If the force increment is being sought, the displacement increment with respect to the state relative to which the displacements are considered, should be realized. In other words, control with respect to force (even during the contact) is being realized by controlling the position, whereby eventual measurement of the force enables finding the grid position, to which the measured force corresponds. From the standpoint of technical realization the problem of accurate controlling of the microdisplacements will arise, which are usually in the hysteresis width of the positional regulation loops. Demonstration of such work are the position oscillations within the scope of the hysteresis width, and accordingly the force oscillations. All the above imposes the need for drives of extremely high quality. In order to avoid this, it is suitable to have in expression (6.6) members with a large displacement, so that their influence in the calculations of force is dominant. That can be realized if the grid is composed of a very rigid part and a part that is very elastic.

During the motion of the cooperative system the derivatives of coordinates are different from zero, so there will appear forces which depend on these derivatives. That is, the change of the kinetic energy δT cannot be excluded from the considerations, as is the case with static conditions, and, if it exists, the dissipation is dependent of velocity. For a consistent description of the motion behavior of the cooperative system, there must be made a correct set of assumptions regarding the elastic properties, either of a part, or of the whole cooperative system, as well as for the properties of the manipulator tip and object contact, as they have been made for static conditions (rigid, pivot, ball, point/surface, with or without friction). The adopted set of assumptions determines the geometrical conditions for determining the statically/dynamically unknown quantities and hence a classification of the cooperative tasks is carried out.

Depending on the assumptions adopted, more or less complex theoretical expressions for the kinetic energy T and deformation work U (potential energy) will be obtained. Based on these, the kinetic potential (Lagrange's function) $L=T-U$ is formed. It is important to emphasize that during motion of the cooperative system, all the demands posed by the system elastic properties have to be satisfied and all basic principles for the motion of the mechanical system have to remain valid. Hence the equations of motion have to be obtained by using some fundamental variational principle (e.g., Hamilton's integral principle or D' Alambert's differential principle) in the form of Lagrange's, Newton's, or Hamilton's equations. The elastic forces in the obtained equations of motion are, according to the principles of Castigliano, the derivatives of deformation work with respect to displacements, and for each instant of the motion of the elastic system attacked by resulting external, inertial, and other loads, they are completely determined from potential energy minimum principle.

Without a sufficiently clear and precise explanation of the previously described physical specificity, and without admitting the necessity of introducing the elastic properties of the cooperative system, but along with research intuition, the elastic properties of the manipulator and/or object have been considered in several papers [14–19]. The models were derived for simple examples and for cases of elastic system motion about the fixed unloaded state of the elastic cooperative system [14–18]. In [19], the general motion of the cooperative system was analyzed, but as a result a description of motion was obtained with twice more state coordinates than necessary.

So, if the assumption about the rigid object, rigid manipulators, and their nonelastic contact is abandoned, the kinematic indefiniteness can appear on manipulators only. The position vectors of contact points in case of elastic contact become independent quantities, so we can only discuss redundant fingers (manipulators), i.e. the relation of the manipulator DOFs number and the working task DOFs number. If the kinematic indefiniteness of manipulator appears instead, the most suitable way of overcoming this is in a way similar to the one nature has given to living beings.

If possible, manipulators should be chosen with the number of DOFs equal to that of the working task. Then, there exists the unique relation between manipulator tips position and attitude, and its derivatives, on one side, and internal coordinates and its derivatives, on the other. If the choice of manipulator DOFs number is not possible, the kinematic chain should be decomposed into a number of parts, equal to the working task DOFs number. For each decomposed part a working space should be found in which a unique relation between the decomposed part position and the points of working space exists. In that space, the most suitable point or region of the decomposed part tip should be chosen using some criterion. Further on, the decomposed subspaces should be regarded as links of a fictitious manipulator, resolving in that manner kinematic relations for given subspaces. Then, within boundaries of given subspace, the relations between subspace coordinates and internal coordinates of links found in this subspace are made. If the movements of the links are small, their configuration should be set such that they form a smooth curve. The analysis of possible ways for resolving the manipulator kinematic indefiniteness is beyond the scope of this consideration.

2. MATHEMATICAL MODEL OF MULTI-ARM COOPERATING ROBOTS

2.1 Introductory Remarks

In order to derive the cooperative manipulation model it is essential to correctly model the motion of the system of bodies which are mutually interconnected in an elastic way (in the sequel: elastic system), and exposed to a system of external forces. Hence, the model of the elastic system is going to be presented first.

First, the static conditions of the elastic system for the zero state (0) are defined, i.e. for the geometrical figure formed under the condition that the contact forces are zero, wherefrom it follows also that the gravitational forces are equal to zero. For these conditions, all displacements of elastic system are also equal to zero. System of forces introduced into the MC generates system motion, generally consisting of translation, rotation, and elastic deformation. Only the motions during which the distances between the points of the elastic system are varying are considered. In other words, system rotation and translation are excluded from considerations. The assumption of the elastic system composed of $m+1$ rigid bodies, mutually interconnected in an elastic way, is introduced. It means that the dimensions of the individual bodies and of the whole system are such, that they do not disturb the clause of the continuity of the elastic hypersurface first derivative, i.e. that the part of the continuous smooth elastic hypersurface (for linear bodies, pat of the elastic line), can be replaced by a hypertendon, sufficiently close to the hypertangent of the elastic hypersurface. Displacements of the elastic system due to deformations, with respect to state 0, form the elastic hypersurface. To each point of the hypersurface correspond the deflexion and the deflexion angle of the hypertangent. For the assumptions introduced, deflexion of the elastic system at the MC of the concrete object corresponds to the translation of that object MC, while the tilts of the elastic hypersurface tangent at point MC correspond to the rotation of that object with respect to state 0, expressed by angular displacements of the coordinate frames located at the MC of the objects, with their axes directed along the main inertia axes. If the elastic system is deformed under static conditions and is found in some arbitrary position outside state 0, then some system of forces must act upon it in order to bring the elastic force into equilibrium. If the system has elastic properties only (i.e. if the mass characteristics are neglected), then the elastic forces are balanced by the external contact forces, i.e. the contact forces are equivalent to the elastic forces produced.

Let the elastic system be slowly taken out of state 0, without translation or rotation of the geometrical figure, and let the corresponding displacements be (Fig. 6.2):

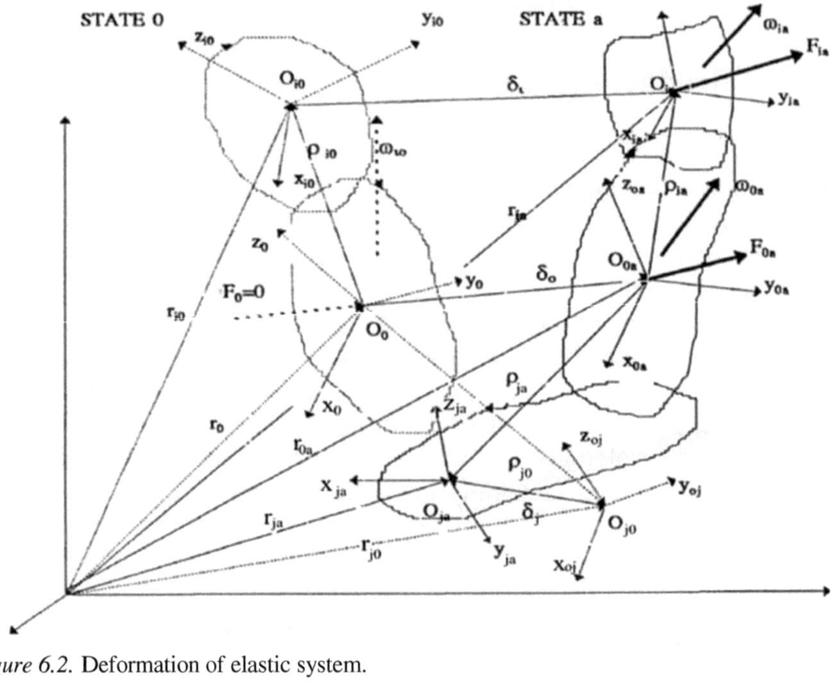

Figure 6.2. Deformation of elastic system.

$$y_i = \begin{pmatrix} \delta_i \\ \mathcal{A}_i \end{pmatrix} = \begin{pmatrix} r_{ia} - r_{i0} \\ \mathcal{A}_{ia} - \mathcal{A}_{i0} \end{pmatrix} \in R^{6\times1}, \quad i=0,1,\ldots,\ m \tag{6.7}$$

where: r_{ia} and r_{i0} are the position vectors of the momentary MC position and the MC position in the state of the $i-$th body part in the 3D Cartesian space; $\mathcal{A}_{ia}$ and $\mathcal{A}_{i0}$ are the vectors of the momentary state of orientation and the orientation relative to that in state 0 of the $i-$th body, measured by the tilt with respect to the Cartesian coordinate system, with origin at MC and its axes directed along the main inertia axes. Index $i=0$ corresponds to a stiff object being manipulated by the manipulators, while the indexes $i=1,\dots,m$ correspond to the connections. According to the elasticity theory, between those displacements, being in the region of linear dependence of stress and dilatation, and the generalized external (in this case, contact) forces/moments $F_i \in R^{6\times 1}$ at points O_{ai}, $i=0,1,\dots,m$, a linear dependence exists, represented in a unique way [11–13]

$$F = Ky \tag{6.8}$$

where $F=(F_0^T F_1^T \dots F_m^T)^T = (F_0^T F_c^T)^T \in R^{6(m+1)\times 1}$ is the expanded vector of generalized forces (forces and moments), $y=(y_0^T y_1^T \dots y_m^T)^T = (y_0^T y_c^T)^T \in$ $\in R^{6(m+1)\times 1}$ is the expanded displacement vector, $K \in R^{6(m+1)\times 6(m+1)}$ is the stiffness matrix, y_0 and F_0 are displacement vectors and the force at the MC of the object being manipulated, and y_c and F_c are the expanded displacement and force vectors at the MC of the connections, respectively. Relation (6.8) comes, by applying the displacement method, from the principle of potential energy minimum when varying the displacement (i.e. the principle of virtual displacements of an elastic body and the statements of Castigliano). The principle states that from all possible displacements permitted by the conditions on the external surface, in reality occur only those for which the system potential energy is at its minimum. Rank of the stiffness matrix can be rank $K \le 6m$ at most, i.e. such an elastic system is kinematically unstable (movable), therefore $6(m+1) - \text{rank } K$ displacements have to be adopted as known (supports displacements or displacements of guides), in order that the system position in the space be determined in a unique way, as a function of these displacements and the acting forces.

Let A be the part of matrix K with rank A = rank K. Then (6.8) can be transformed into

$$\begin{pmatrix} F_a \\ F_v \end{pmatrix} = \begin{pmatrix} A & b \\ c & d \end{pmatrix} \cdot \begin{pmatrix} y_a \\ y_v \end{pmatrix} \Rightarrow y_a = A^{-1} F_a - A^{-1} b y_v \tag{6.9}$$

where F_a is the vector made of chosen forces rank K, y_a are the corresponding displacements produced where forces F_a act, F_v and y_v are the remaining forces and displacements, and b, c, d are the corresponding vectors of the reduced matrix.

From the preceding text the following conclusions can be drawn.

In order to determine the forces at contact it is essential to suppose the elasticity of connections. For force calculations it is necessary to know the characteristics of the elastic connections, and to know the displacements of the connections and the object MC with respect to state 0 (state of making contact). The values of the displacement of knots can be obtained by measurement or by calculation.

2.2 Elastic System Model for Unloaded State at Rest (State 0)

Let the bodies have masses and the elastic system be moving about the state at rest 0. Motion equations of that system will be derived as second-order Lagrange's equations.

Displacements of the MC with respect to the unloaded state y_i, defined by (6.7), are adopted as generalized coordinates. Potential energy Π of the elastic system is equal to the deformation work

$$2\Pi = y^T K y,\ K = K^T \in R^{6(m+1)\times 6(m+1)},\ \text{rank}\, K \leq 6m \tag{6.10}$$

Let us define the elastic forces as forces that are to act in the knots of the elastic grid, in order that it could find itself in a predetermined position in space. The elastic forces are equal to the derivatives of the deformation work with respect to displacements $F_e = \partial\Pi/\partial y = Ky \in R^{6(m+1)\times 1}$, $F_{ei} = \partial\Pi/\partial y_i = K_i\, y \in R^{6\times 1}$, $K_i \in R^{6\times 6(m+1)}$ where K_i are the submatrices, composed of $6i+1$ to $6i+6-$th rows of matrix K.

Relation between the angular velocity $\omega_i = (p_i q_i r_i)^T$ measured along the $i-$th body main inertia axes and the derivatives of orientation given by $\dot{\mathcal{A}}_i = (\dot{\psi}\dot{\theta}\dot{\varphi})^T$, is given by $\omega_i = (p_i q_i r_i)^T = L_\omega(\mathcal{A}_i)\cdot\dot{\mathcal{A}}_i$. Motion velocity $v_{ia} = col(\dot{r}_{ia}, \omega_{ia}) = diag(I_{3x3}, L_\omega(\mathcal{A}_i))\, col(\dot{\delta}_i, \dot{\mathcal{A}}_i) = L_v(\mathcal{A}_i)\dot{y}_i = L_v(y_i)\dot{y}_i$. Total kinetic energy is defined by $T = T_0 + T_1 + \ldots + T_i + \ldots + T_m$, while the kinetic energy of the $i-$th part is given by $2T_i = m_i\,\dot{\delta}_i^T\,\dot{\delta}_i + \omega_i^T I_i\,\omega_i = m_i\,\dot{\delta}_i^{x^2} + m_i\,\dot{\delta}_i^{y^2} + m_i\,\dot{\delta}_i^{z^2} + A\dot{p}_i^2 + B\dot{q}_i^2 + C\dot{r}_i^2$, where m_i is the mass and A_i, B_i, C_i are the i-th object main moments of inertia. By substituting the angular velocity one obtains:

$2T_i = m_i \dot{\delta}_i^T \dot{\delta}_i + \dot{\mathcal{A}}_i^T L_\omega^T(\mathcal{A}_i) I_i L_\omega(\mathcal{A}_i)\dot{\mathcal{A}}_i = \dot{y}_i^T L_v^T(y_i)\, M_i L_v(y_i)\dot{y}_i = \dot{y}_i^T W_i(y_i)\dot{y}_i$,
where $W_i(y_i) = L_v^T(y_i) M_i L_v(y_i) \in R^{6\times 6}$, $W_i(y_i) = W_i^T(y_i)$, $\det W_i(y_i) \neq 0$, $I_i = diag(A_i, B_i, C_i)$ and $M_i = (m_i, m_i, m_i, A_i, B_i, C_i)$.

Derivatives of kinetic energy with respect to velocity are defined by $\partial T_i / \partial \dot{y}_i = L_v^T(y_i) M_i L_v(y_i)\dot{y}_i = W_i(y_i)\dot{y}_i$ and $d/dt(\partial T_i / \partial \dot{y}_i) = \dot{W}_i(y_i)\dot{y}_i + W_i(y_i)\ddot{y}_i$.

The preceding equations for kinetic energy can be given briefly as

$$\begin{aligned} 2T = \sum_{i=0}^{m} \left(\sum_{j=1}^{\infty} dm_j v_j\right) = \sum_{i=0}^{m} m_i \dot{\delta}_i^2 + \sum_{i=0}^{m} I_i \omega_i^2 = \\ = \dot{y}^T L_v^T(y) M L_v(y)\dot{y} = \dot{y}^T W(y)\dot{y} \end{aligned} \tag{6.11}$$

where

$M = diag(M_o, M_1, \dots M_n) \in R^{6(m+1)\times 6(m+1)}$,
$M_i = diag(m_i, m_i, m_i, A_i, B_i, C_i) \in R^{6\times 6}$, $v = (v_o^T v_1^T \dots v_m^T)^T \in R^{6(m+1)\times 1}$,
$v_i = (\dot{\delta}_i^T \omega_i^T(\mathcal{A}_i))^T = L_{vi}(y_i)\dot{y}_i \in R^{6\times 1}$, $\omega_i = L_{\omega i}(\mathcal{A}_i)\dot{\mathcal{A}}_i \in R^{3\times 1}$,
$L_{vi}(y_i) = diag(I_{3\times 3}, L_\omega(\mathcal{A}_i)) \in R^{6\times 6}$, $L_v(y) = diag(L_{v0}, L_{v1}, \dots L_{vm}) \in$
$\in R^{6(m+1)\times 6(m+1)}$

If the connections have dissipative properties, Rayleigh's function (dissipation energy) and derivative of the dissipation energy with respect to velocity can be expressed as

$$2\mathcal{D} = -\dot{y}^T D\dot{y}, \qquad \frac{\partial \mathcal{D}}{\partial \dot{y}} = -D\dot{y}, \qquad \frac{\partial \mathcal{D}}{\partial \dot{y}_i} = -D_i\,\dot{y} \tag{6.12}$$

where $D = D^T \geq 0$, $D \in R^{6(m+1)\times 6(m+1)}$ is a matrix with damping coefficients corresponding to velocities, and $D_i \in R^{6\times 6(m+1)}$ are submatrices composed of $6i+1$ to $6(i+1)$–th rows of matrix D, inclusive.

By substituting these expressions into Lagrange's equations we obtain

$$\frac{d}{dt}\frac{\partial T}{\partial \dot{y}_i} - \frac{\partial T}{\partial y_i} - \frac{\partial \mathcal{D}}{\partial \dot{y}_i} + \frac{\partial \Pi}{\partial y_i} = Q_i, \qquad i = 0, 1, \dots, m \tag{6.13}$$

where
$Q_i=(Q_i^1 ... Q_i^6)^T = G_i(m_i g)+F_i$, $y_i=(y_i^1 y_i^2 y_i^3 y_i^4 y_i^5 y_i^6)^T=(\delta_i^x \delta_i^y \delta_i^z \psi_i \theta_i \varphi_i)^T$.
For the elastic system performing general motion about the unloaded state at rest 0, under the action of external forces F, a general form of model is obtained as

$$W_i(y_i)\ddot{y}_i + \dot{W}_i(y_i)\dot{y}_i - \frac{\partial T_i}{\partial y_i}(y_i, \dot{y}_i) + D_i\dot{y} + K_i y =$$

$$= G_i(m_i g) + F_i, \qquad i=0,1,...,m$$

or in condensed form

$$W_i(y_i)\ddot{y}_i + w_i(y, \dot{y}) = F_i, \qquad i=0,1,...,m \tag{6.14}$$

where

$$w_i(y, \dot{y}) = \dot{W}_i(y_i)\dot{y}_i - \partial T_i(y_i, \dot{y}_i)/\partial y_i + D_i\dot{y} + K_i y - G_i(m_i g) \in R^{6\times 1}, \quad i=0,1,...,m$$

By uniting all $6(m+1)$ equations it is obtained that

$$W(y)\ddot{y} + w(y, \dot{y}) = F \tag{6.15}$$

where

$W(y) = diag(W_0(y_0)\ W_1(y_1) ... W_m(y_m)) \in R^{6(m+1)\times 6(m+1)}$, $W(y) = W^T(y)$, $\det W(y) \neq 0$, $w(y, \dot{y}) = (w_0^T(y, \dot{y}) ...\ w_m^T(y, \dot{y}))^T \in R^{6(m+1)\times 1}$.

From $6(m+1)$ equations (6.15), the number of independent equations is exactly equal to the stiffness matrix rank (rank K). Equation (6.15) can be presented in such a way that the descriptions of connections motion and manipulation object are separated

$$W_c(y_c)\ddot{y}_c + w_c(y, \dot{y}) = F_c$$

$$W_0(y_0)\ddot{y}_0 + w_0(y, \dot{y}) = 0 \tag{6.15a}$$

where the index c designates the quantities connected to the contact points, and index 0 designates the quantities connected to the manipulated object.

Hereby are: $y_c=(y_1^T y_2^T ... y_m^T)^T \in R^{6m\times1}$, $y_0 \in R^{6\times1}$, $F_c=(F_1^T F_2^T ... F_m^T)^T \in R^{6m\times1}$, $F_0=0 \in R^{6\times1}$, $W_c(y_c)=diag(W_1(y_1)...W_m(y_m)) \in R^{6m\times6m}$, $W_c(y_c)=W_c^T(y_c)$, $\det W_c(y_c) \neq 0$, $w_c(y,\dot{y})=(w_1^T(y,\dot{y})...w_m^T(y,\dot{y}))^T \in R^{6m\times1}$, where y_c denotes the expanded contact position vector in $6m$ dimensional space and F_c the expanded vector of the contact forces, adjoined to that vector. Let us mention that on the manipulated object MC no contact force acts directly, hence $F_0=0$. Equations (6.15) and (6.15a) represent the final form of the equations of the behavior of the elastic system which, under the action of external contact forces F_c, performs general motion about the unloaded state 0 at rest.

2.3 Model of Elastic System for Moving Unloaded State

Let the geometrical figure with state 0 move through space. Stress of the elastic system takes place in the same way as when state 0 is at rest. In other words, stress of the elastic system is still regarded only with respect to the moving state 0, while this motion of state 0 influences only the members which do not reflect the elastic system elastic properties. Simply, the potential energy and the dissipation energy of the elastic system connections are determined by displacement of the system relative to the moving unloaded state, while the other quantities are defined for the elastic system absolute coordinates.

If it is supposed that the position of the loaded state is known, then, in order to determine the displacements with respect to state 0, the position of state 0 should be determined. Position of any point of the loaded elastic system can be described in Cartesian space by defining displacement with respect to the unloaded state 0, or by the coordinates of the initial unloaded state which, for the loaded state, represent curvilinear coordinates. It is possible to express the coordinates of any point of the initial unloaded state as functions of the position coordinates of that point in loaded state. Then, the Cartesian coordinates of the loaded state are curvilinear coordinates for the unloaded state. In other words, if the connected system $Oxyz$ and system $O_{i0}x_{i0}y_{i0}z_{i0}$ describing state 0 are adopted as Cartesian, then the absolute coordinates of any point of the loaded (stressed) elastic system are curvilinear coordinates of that point in the unloaded state, and vice versa. If the absolute coordinates of any point of the loaded elastic system are adopted as Cartesian, then these coordinates in the unloaded state 0 are curvilinear coordinates of the absolute coordinates of that point [20]. Solution of the problem for known shape, dimensions, material characteristics, and acting forces for the ideal elastic system involves finding fifteen functions; six component of stress tensor, six component of dilatation tensor, and three component of displacements. Exact solution methods reduce to solving Lame's equations in which only displacements figure, or the Beltrami–Mitchell's equations which define the state of stress, or by the semi-inverse Saint Venant's method. Displacement methods and force methods are known from the variation calculus. In engineering practice approximate methods are widely applied, which can be divided into two groups [11–13]. The first group includes the methods based on approximate solutions of the differential equations system of an approximate model with a finite number of DOFs. Some of these methods are: the method of finite differences, the method of iteration, and the finite elements method. The second group, represented by the variational methods, are the known Rayleigh–Ritz's method and Galerkin's method.

The idea of the introduced assumptions and of the proposed modeling procedure is to avoid solving systems of equations describing the elastic system deformations, and deriving instead the model based only on knowing the mass centers absolute coordinates, their derivatives and the reference points positions in the initial moment, i.e. the positions of the contact points and the manipulated object MC (Fig. 6.1). Here, the modelling procedure for the elastic system in the form of linear and torsional springs placed between any two of its knots is presented. Such choice enables the relative displacements of the deformation work of the elastic system to be directly determined as functions of the absolute coordinates.

State 0 can be represented by one coordinate system only and it is adopted that this system is $O_0x_0y_0z_0$, the orientation of which in the considered instant is given by the vector $A^0 = (\psi^0\theta^0\phi^0)^T$ and by the position of the coordinates origin r_0. The other coordinate system $O_{i0}x_{i0}y_{i0}z_{i0}$ is deflected with respect to the system $O_0x_0y_0z_0$ by some constant orientation value $\mathcal{A}_{i0}$ during all the time of motion. It is implicitly understood that this consideration is carried out relative to orientation $\mathcal{A}^0, \mathcal{A}^0 + \mathcal{A}_{10}, \ldots, \mathcal{A}^0 + \mathcal{A}_{m0}$, and for more clarity it is adopted that $\mathcal{A}_{i0} = 0$, $i = 1, \ldots, m$.

According to what has been discussed and Fig. 6.2 we can form the vector functions

$$r_{ia} = r_{i0} + \delta_i\,,\ \ \dot{r}_{ia} = \dot{r}_{i0} + \dot{\delta}_i\,,\quad i = 0,1,\ldots,m \tag{6.16}$$

where r_{i0} is the MC position of the $i-$th body in unloaded state

$$r_{i0} = r_0 + \rho_{i0}\,,\ \dot{r}_{i0} = \dot{r}_0 + \dot{\rho}_{i0} \tag{6.17}$$

where ρ_{i0} is the position vector of the $i-$th body MC in the unloaded state 0 of the elastic system, given in frame $O_0x_0y_0z_0$.

State 0 is given by

$$r_0 = r_{ia} - \rho_{i0} - \delta_i, \quad \dot{r}_{ia} = \dot{r}_{i0} + \dot{\delta}_i, \qquad i = 0,1,\dots,m$$

$$\mathcal{A}^0 = \mathcal{A}_{ia} - \mathcal{A}_i,\ \dot{\mathcal{A}}^0 = \dot{\mathcal{A}}_{ia} - \dot{\mathcal{A}}_i, \qquad i = 0,\ 1,\dots,m \tag{6.18}$$

Angular displacement between two arbitrary bodies is defined by the difference of the absolute values of their orientation. Namely, since before the elastic system deformation the bodies possess the same orientation $\mathcal{A}^0$, or their orientations differ by a constant value $\mathcal{A}^0_{ij}$, then by three (or, if $\mathcal{A}^0_{ij} \neq 0$ by four) successive alterations of orientation is possible to pass from the state with the orientation $\mathcal{A}_{i0}$ to that same state. Let the initial state have orientation $\mathcal{A}_{i0}$. By the orientation change $\mathcal{A}_{ia} - \mathcal{A}_{i0}$, it is arrived at a state of orientation $\mathcal{A}_{ia} = \mathcal{A}_{i0} + (\mathcal{A}_{ia} - \mathcal{A}_{i0})$. By the orientation change $-\mathcal{A}_{ija} = -(\mathcal{A}_{ia} - \mathcal{A}_{ja})$, it is arrived at a state of orientation $\mathcal{A}_{ja} = \mathcal{A}_{i0} + (\mathcal{A}_{ia} - \mathcal{A}_{i0}) - (\mathcal{A}_{ia} - \mathcal{A}_{ja})$. By further orientation changes by $-(\mathcal{A}_{ja} - \mathcal{A}_{j0})$ and $(\mathcal{A}_{i0} - \mathcal{A}_{j0})$, one arrives at the state $\mathcal{A}_{j0} = \mathcal{A}_{i0} + (\mathcal{A}_{ia} - \mathcal{A}_{i0}) - (\mathcal{A}_{ia} - \mathcal{A}_{ja}) - (\mathcal{A}_{ja} - \mathcal{A}_{j0})$ and $\mathcal{A}_{end} = \mathcal{A}_{i0} + (\mathcal{A}_{ia} - \mathcal{A}_{i0}) - (\mathcal{A}_{ia} - \mathcal{A}_{ja}) - (\mathcal{A}_{ia} - \mathcal{A}_{ja}) - (\mathcal{A}_{ja} - \mathcal{A}_{j0}) + (\mathcal{A}_{i0} - \mathcal{A}_{j0})$, respectively. By simple addition it is evident that $\mathcal{A}_{end} = \mathcal{A}_{i0}$, i.e. we have returned to the state with the initial orientation value. It follows that the mutual orientation between two arbitrary bodies in loaded state is defined by the difference of the absolute values of their orientation $\mathcal{A}_{ij} = \mathcal{A}_{ija} = \mathcal{A}_{ia} - \mathcal{A}_{ja}$. According to the preceding, the mutual displacements of the loaded system points are defined by the difference of the displacements of points with respect to the unloaded state

$$\delta_{ij} = -\delta_{ji} = \delta_i - \delta_j = r_{ia} - r_{ja} - (r_{i0} - r_{j0}) = \rho_{ija} - \rho_{ij0},$$

$$\mathcal{A}_{ij} = -\mathcal{A}_{ji} = \mathcal{A}_{ia} - \mathcal{A}_{ja}, \qquad i,\ j = 0\dots m \tag{6.19}$$

Velocity of the loaded system points displacement with respect to the unloaded state is

$$\dot{\delta}_i = \dot{r}_{ia} - \dot{r}_{i0} = \dot{r}_{ia} - \dot{r}_0 - \omega_0 \times \rho_{i0} \tag{6.20}$$

where ω_0 is angular velocity of the geometric figure formed by the contact points of the elastic system in the state 0.

As no force system acts in the unloaded state, state 0 moves as a rigid body, so

$$\| \rho_{ij0} \| = \| \rho_{i0} - \rho_{j0} \| = \text{const},$$

$$\dot{\rho}_{ij0} = \omega_0 \times \rho_{ij0} = -\rho_{ij0} \times \omega_0 \tag{6.21}$$

i.e. the vector ρ_{ij0} possesses a constant intensity and a variable direction. In the case of translatory motion of the unloaded state 0, $\omega_0 = 0$ and $\rho_{ij0} =$ const, and in the case of coinciding of the MC position of $i-$th and $j-$th body, in state 0, $\rho_{ij0} = 0$.

Thereby, velocity of change of the mutual distance of the loaded state point is

$$\dot{\delta}_{ij} = \dot{\delta}_i - \dot{\delta}_j = \dot{r}_{ia} - \dot{r}_{ja} - \dot{\rho}_{ij0} = \dot{r}_{ia} - \dot{r}_{ja} - \omega_0 \times \rho_{ij0},$$

$$\dot{\mathcal{A}}_{ij} = \dot{\mathcal{A}}_{ia} - \dot{\mathcal{A}}_{ja} \tag{6.22}$$

By uniting (6.19) and (6.22) the relation of displacement and velocity of displacement between the points of arbitrary MC connections and the manipulated object is obtained

$$y_{ij} = \begin{pmatrix} \delta_{ij} \\ \mathcal{A}_{ij} \end{pmatrix} = \begin{pmatrix} r_{ia} - r_{ja} - \rho_{ij0} \\ \mathcal{A}_{ia} - \mathcal{A}_{ja} \end{pmatrix} \tag{6.23}$$

$$\dot{y}_{ij} = \begin{pmatrix} \dot{\delta}_{ij} \\ \dot{\mathcal{A}}_{ij} \end{pmatrix} = \begin{pmatrix} \dot{r}_{ia} - \dot{r}_{ja} + \rho_{ij0} \times \omega_0 \\ \dot{\mathcal{A}}_{ia} - \dot{\mathcal{A}}_{ja} \end{pmatrix} \tag{6.24}$$

In order to present the elastic system general model in the form of Lagrange's equations, it is necessary to define the kinetic, potential, and dissipation energies. The absolute coordinates $Y_i = (r_{ia}^T \; \mathcal{A}_{ia}^T)^T \in R^{6\times 1}$, $i = 0,\ldots,m$ of the mass centers for the loaded elastic system state are adopted as generalized coordinates. Kinetic energy is defined by the absolute velocities as

$$2T_a = \sum_{i=0}^{m}(\sum_{j=1}^{\infty} dm_j v_j) = \sum_{i=0}^{m} m_i \dot{r}_{ia}^2 + \sum_{i=0}^{m} I_i \omega_{ia}^2 = \frac{1}{2}\sum_{0}^{m} \dot{Y}_i^T L_{via}(\dot{Y}_i) = \qquad (6.25)$$
$$= \dot{Y}^T L_{va}^T(Y) M L_{va}(Y)\dot{Y} = \dot{Y}^T W_a(Y)\dot{Y}$$

where

$Y_i = (r_{ia}^T \mathcal{A}_{ia}^T)^T \in R^{6\times1}$, $Y = (Y_0^T Y_1^T \ldots Y_m^T)^T \in R^{6\,(m+1)\times1}$, $v_i = (\dot{r}_{ia}^T \omega_{ia}^T(\mathcal{A}_{ia}))^T =$
$= L_{via}(Y_i)\dot{Y}_i \in R^{6\times1}$, $\omega_{ia} = L_{\omega ia}(\mathcal{A}_{ia})\dot{\mathcal{A}}_{ia} \in R^{3\times1}$, $v_a = (v_{0a}^T\; v_{1a}^T \ldots v_{ma}^T)^T \in R^{6(m+1)\times1}$,
$L_{via}(Y_i) = diag(I_{3\times3}, L_{\omega ia}(\mathcal{A}_{ia})) \in R^{6\times6}$, $L_{va}(Y) = diag(L_{v0a}, L_{v1a} \ldots L_{vma}) \in R^{6(m+1)\times6(m+1)}$,
$W_a(Y) = diag(W_{0a}, W_{1a}, \ldots, W_{ma}) = L_{va}^T(Y) M L_{va}(Y) \in R^{6(m+1)\times6(m+1)}$.

Form W_{ia} is identical to form W_i, whereby instead of index i, index ia should be used, as designation of absolute coordinate. Hence the derivatives of the kinetic energy

$$\frac{\partial T_{ia}}{\partial \dot{Y}_i} = W_{ia}(Y_i)\dot{Y}_i\,,\; \frac{d}{dt}\frac{\partial T_{ia}}{\partial \dot{Y}_i} = \dot{W}_{ia}(Y_i)\dot{Y}_i + W_{ia}(Y_i)\ddot{Y}_i\,,$$
$$\frac{\partial T_{ia}}{\partial Y_i}(Y_i, \dot{Y}_i) = \frac{1}{2}\sum_{i=0}^{m} \dot{Y}_i^T \frac{\partial L_{via}(Y_i)}{\partial Y_i}\dot{Y}_i \qquad (6.26)$$

are also identical to the expressions obtained in (6.22), in which the indexes should be changed, i.e. instead of the displacement coordinates with index i, the absolute coordinates with index ia should be used.

It was already stated that the elastic system potential energy is equal to the elastic system deformation work. For an idealy elastic system, work does not depend on the characteristic of the trajectory along which the system passes from one state to the other, but depends only on the scalar product of the system displacements and forces. This enables the projection of vectors either onto the force direction, or onto the displacement direction. Vectors of mutual displacement of points i and j are defined by (6.23). However, in ρ_{ij0} the unknown orientation $\mathcal{A}^0$ of state 0 figures, because y_{ij} defines the increment (rotation and deformation) of the vector during stress of the elastic system. The angular displacement has already been defined by (6.23), therefore, to define the work it is still necessary to define the extension y_{ij}^D of vector ρ_{ij0} in the course of deformation, in which the orientation $\mathcal{A}^0$ does not figure. Since the intensity $\|\rho_{ij0}\| =$ const does not change, the linear displacement of points i and j and the orientation change of the axes, along which the work is being performed, are defined by

$$y_{ij}^D = y_{ij}^D(Y) = \begin{pmatrix} \rho_{ija} - \left\| \rho_{ij0} \right\| \dfrac{\rho_{ija}}{\left\| \rho_{ija} \right\|} \\ \mathcal{A}_{ia} - \mathcal{A}_{ja} \end{pmatrix} =$$

$$= \begin{pmatrix} 1 - \dfrac{\left\| \rho_{ij0} \right\|}{\left\| r_{ia} - r_{ja} \right\|} (r_{ia} - r_{ja}) \\ \\ \mathcal{A}_{ia} - \mathcal{A}_{ja} \end{pmatrix} = \Lambda_{ij}\ (Y_i, Y_j)\ (Y_i - Y_j) \tag{6.27}$$

where

$\Lambda_{ij}\ (Y_i, Y_j)\ (Y_i - Y_j) = diag((1 - (\left\| \rho_{ij0} \right\| / \left\| r_{ia} - r_{ja} \right\|))\, I_{3\times3}, I_{3\times3}) \in R^{6\times6}$.

Total potential energy due to linear and rotational displacement of body i relative to body j,i,j=0,1,...m is defined by $\Pi_a = \Pi_{01a} + \Pi_{02a} + \Pi_{03a} + ... + \Pi_{0ma} + \Pi_{12a} + \Pi_{13a} + ... + \Pi_{1ma} + ... + \Pi_{(m-1)ma}$. Arbitrary member Π_{ija} of this sum is defined by $2\Pi_{ija} = (y_{ij}^D)^T K_{ija}\, y_{ij}^D = (Y_i - Y_j)^T \Lambda_{ij}(Y_i, Y_j)\, K_{ija} \times \Lambda_{ij}(Y_i, Y_j)(Y_i - Y_j) = Y_i^T \pi_{ij} Y_i - 2Y_i^T \pi_{ij} Y_j + Y_j^T \pi_{ij} Y_j$, where $\pi_{ij} = \pi_{ji} = \Lambda_{ij}(Y_i, Y_j) K_{ija} \Lambda_{ij} = diag(c_{ij}\,(1 - (\left\| \rho_{ij0} \right\| / \left\| r_{ia} - r_{ja} \right\|))\, I_{3\times3}, I_{3\times3}) \in R^{6\times6}$, whereby, after substitution, total potential energy is determined $2\Pi_a = Y_0^T(\pi_{01} + ... + \pi_{0m})Y_0 - Y_0^T \pi_{01} Y_1 - ... - Y_0^T \pi_{0m} Y_m ... - Y_m^T \pi_{m0} Y_0 - ... + Y_m^T(\pi_{m0} + ... + \pi_{m(m-1)})\, Y_m = Y^T \pi_a(Y)\, Y$ or in the matrix form

$$2\Pi_a = Y^T \pi_a(Y)\, Y \tag{6.28}$$

Elastic forces are equal to the derivative of potential energy with respect to the coordinates and defined by

$$F_e = \frac{\partial \Pi_a}{\partial Y} = \frac{1}{2} \frac{\partial Y^T \bar{\pi}_a Y}{\partial Y} + \pi_a(Y)\, Y \in R^{6(m+1)\times1},$$

$$F_{ei} = \frac{\partial \Pi_a}{\partial Y_i} = \frac{1}{2} \frac{\partial Y^T \bar{\pi}_a Y}{\partial Y_i} + \pi_{ia}(Y)\, Y \in R^{6\times1},$$

where $\pi_{ia}(Y) \in R^{6\times 6(m+1)}$ are submatrices composed of $6i+1$ to $6(i+1)$-th row of matrix $\pi_a(Y)$, and $\partial(Y^T \overline{\pi}_a Y)/\partial Y_i$ is the vector of the quadratic form (scalar) derivative $Y^T \pi_a Y$ with respect to vector Y_i, whereby the superscript dash designates that partial derivation was carried out over the matrix π_a.

Attention has been drawn to the fact that between the internal and elastic forces there exists a unique relationship [11–13]. When F_e is introduced into this relation, the dependence of the internal forces on the displacement or the absolute coordinates of knots of the elastic grid is obtained. Hence the realized values of the internal forces are functions of the geometric configuration of the elastic system.

We have to proceed in an analogous way to determine the dissipation energy. Displacement of the i-th object with regard to the j-th one is composed of a translation along vector ρ_{ija} and one mutual rotation. The resulting vectors of these two motions are normal. Energy losses proportional to velocity can be produced only along the trajectory during these motions. The translatory component $\dot{\delta}_{ij}^D$ of the displacement velocity $\dot{\delta}_{ij}$ equals the projection of this velocity into the direction of vector ρ_{ija}, which is described by

$$\begin{aligned}\dot{\delta}_{ij}^D &= \frac{(\dot{\delta}_{ij}^T \cdot \rho_{ija})}{\|\rho_{ija}\|} \cdot \frac{\rho_{ija}}{\|\rho_{ija}\|} = \\ &= (\dot{\delta}_{ij}^T \cdot (r_{ia} - r_{ja})) \cdot \frac{r_{ia} - r_{ja}}{\|r_{ia} - r_{ja}\|^2} = \frac{(r_{ia} - r_{ja}) \cdot (r_{ia} - r_{ja})}{(r_{ia} - r_{ja})^T \cdot (r_{ia} - r_{ja})} \cdot \dot{\delta}_{ij}^T\end{aligned} \tag{6.29}$$

In the case of co-linearity of $\dot{\delta}_{ij}$ and ρ_{ija}, $\dot{\delta}_{ij}^R = \rho_{ij0} \times \omega_0 = 0$ and $(\dot{\delta}_{ij}^T \cdot \rho_{ija}) / \|\rho_{ija}\| = \|\dot{\delta}_{ij}\|$, and as the unit vectors are identical, also, $\dot{\delta}_{ij} / \|\dot{\delta}_{ij}\| = \rho_{ija} / \|\rho_{ija}\|$ it follows $\dot{\delta}_{ij}^D = \dot{\delta}_{ij} = \dot{r}_{ia} - \dot{r}_{ja}$. For small displacements of the elastic system, the vectors $\rho_{ija} = r_{ia} - r_{ja}$ and $\rho_{ij0} \times \omega_0$ are approximately normal, so $(\rho_{ij0} \times \omega_0)^T \cdot (r_{ia} - r_{ja}) \approx 0$, whereby

$$\dot{\delta}_{ij}^D \simeq \frac{(\dot{r}_{ia} - \dot{r}_{ja})^T \cdot (r_{ia} - r_{ja})}{\|r_{ia} - r_{ja}\|^2} \cdot (r_{ia} - r_{ja}) = \mathcal{G}_{ija}(r_{ia}, r_{ja})(\dot{r}_{ia} - \dot{r}_{ja}) \tag{6.30}$$

while the rotational component is given by $\dot{\delta}_{ij}^R = \dot{\delta}_{ij} - \dot{\delta}_{ij}^D =$ $= (I - \mathcal{G}_{ija}(r_{ia}, r_{ja}))(\dot{r}_{ia} - \dot{r}_{ja}) + \rho_{ij0} \times \omega_0$, where $\mathcal{G}_{ija}(r_{ia}, r_{ja}) = \mathcal{G}_{ija}^T(r_{ia}, r_{ja}) =$ $= (r_{ia} - r_{ja}) \cdot (r_{ia} - r_{ja}) / (r_{ia} - r_{ja})^T \cdot (r_{ia} - r_{ja}) \in R^{3\times3}$, det $\mathcal{G}_{ija}(r_{ia}, r_{ja}) =$ $= \Gamma(r_{ia} - r_{ja}) / \left\| r_{ia} - r_{ja} \right\|^2 = 0$, $r_{ia} \neq r_{ja}$, $rank \mathcal{G}_{ija}(r_{ia}, r_{ja}) = 1$, and $\Gamma(r_{ia} - r_{ja})$ is the determinant of matrix $(r_{ia} - r_{ja}) \cdot (r_{ia} - r_{ja})$. In the case of translatory motion of state 0, $\omega_0 = 0$, and if vectors $r_{i0}, r_{j0}, r_{ia}, r_{ja}$, i.e. ρ_{ij0} and ρ_{ija} are co-linear, then $\mathcal{G}_{ija}(r_{ia}, r_{ja})\ (\dot{r}_{ia} - \dot{r}_{ja}) = \left\| \rho_{ij0} \right\| (\dot{r}_{ia} - \dot{r}_{ja})$ and $\dot{\rho}_{ij0} = \rho_{ij0} \times \omega_0 = 0$ whereby $\dot{\delta}_{ij}^R = 0$.

Total dissipation energy consumed in the course of linear and rotational displacement of body i relative to body j, i, $j = 0, 1, ...,\ m$ is defined by $\mathcal{D}_a = \mathcal{D}_{01a} + \mathcal{D}_{02a} + \mathcal{D}_{03a} + ... + \mathcal{D}_{0ma} + \mathcal{D}_{12a} + \mathcal{D}_{13a} + ... + \mathcal{D}_{1ma} + \mathcal{D}_{23a} + ... + \mathcal{D}_{2ma} + ... + \mathcal{D}_{(m-1)ma}$. Arbitrary member $\mathcal{D}_{ija}$ of that sum is defined by $-2\mathcal{D}_{ija} = (\dot{\delta}_{ij}^D)^T D_{ij}^{\delta} \dot{\delta}_{ij}^D +$ $+ (\dot{\mathcal{A}}_{ia} - \dot{\mathcal{A}}_{ja})^T D_{ij}^{\mathcal{A}} (\dot{\mathcal{A}}_{ia} - \dot{\mathcal{A}}_{ja}) = \dot{Y}_i^T D_{ij} \dot{Y}_i - 2\dot{Y}_i^T D_{ij} \dot{Y}_j + \dot{Y}_j^T D_{ij} \dot{Y}_j$ where $D_{ij} = D_{ij}^T = D_{ji} = diag\ \mathcal{G}_{ija}(r_{ija}, r_{ja})\ D_{ij}^{\delta} \mathcal{G}_{ija}(r_{ia}, r_{ja}),\ (D_{ij}^{\mathcal{A}}) \in R^{6\times6}$. Whereby, after substitution, the total dissipation energy is determined by $-2\mathcal{D}_a = \dot{Y}_0^T (D_{01} + ... + D_{0m}) \dot{Y}_0 - \dot{Y}_0^T D_{01} \dot{Y}_1 - ... - \dot{Y}_0^T D_{0m} \dot{Y}_m ... - \dot{Y}_m^T D_{m0} \dot{Y}_0 -$ $... + \dot{Y}_m^T (D_{m0} + D_{m1} + ... + D_{m(m-1)}) \dot{Y}_m = -\dot{Y}^T D_a(Y) \dot{Y}$, due to $D_{ij} = D_{ji}$. In expanded form $\pi_a(Y)$ and $D_a(Y)$ are (sign * stands for D or π):

$$
*_a(Y) = *_a^T(Y) = \begin{bmatrix} \sum_{k=0, k\neq0}^{n} *_{0k} & -*_{01} & -*_{02} & \cdots & -*_{0m} \\ -*_{01} & \sum_{k=0, k\neq1}^{n} *_{1k} & -*_{12} & \cdots & -*_{1m} \\ \cdots & \cdots & \cdots & \cdots & \cdots \\ -*_{0m} & -*_{1m} & -*_{2m} & \cdots & \sum_{k=0, k\neq m}^{n} *_{km} \end{bmatrix} \qquad (6.31)
$$

The derivative with respect to velocity is defined by $\partial \mathcal{D}_a / \partial \dot{Y} = -D_a(Y)\dot{Y}$, $\partial \mathcal{D}_a / \partial \dot{Y}_i = -D_{ia}(Y)\dot{Y} \in R^{6\times1}$, where $D_{ia}(Y) \in R^{6\times6(m+1)}$ are submatrices composed of $6i + 1$ to $6i + 6 -$ th row, inclusively, of matrix

$D_a(Y) \in R^{6(m+1)\times 6(m+1)}$. Substituting the obtained expressions into the Lagrange's equations

$$\frac{d}{dt}\frac{\partial T_a}{\partial \dot{Y}_i^j} - \frac{\partial T_a}{\partial Y_i^j} - \frac{\partial \mathcal{D}_a}{\partial \dot{Y}_i^j} + \frac{\partial \Pi_a}{\partial Y_i^j} = Q_{ia}^j, \quad i=0,1,...,m, \quad j=1,...,6 \quad (6.32)$$

where the generalized forces for the individual components Y_i^j of vector Y_i are given by

$$Q_{ia}^j = \sum_{k=0}^{m} (\partial Y_k / \partial Y_i^j)\,(G_k(m_k g) + F_k) = G_i^j(m_i g) + F_i^j, i = 0,1,...,m, j = 1,...,6$$

for an elastic system performing general motion under the action of a system of external forces about the unloaded state 0, which is also performing general motion, the general form of model is obtained as

$$W_{ia}(Y_i)\ddot{Y}_i + \dot{W}_{ia}(Y_i)\dot{Y}_i - \frac{\partial T_{ia}}{\partial Y_i}(Y_i, \dot{Y}_i) + D_{ia}\dot{Y} + \frac{1}{2}\frac{\partial Y^T \overline{\pi}_a Y}{\partial Y_i} + \pi_{ia}(Y)Y = G_i(m_i\, g) + F_i$$

$i = 0, 1, ..., m,$ or in a abbreviated form

$$W_{ia}(Y_i)\ddot{Y}_i + w_{ia}(Y, \dot{Y}) = F_i, \qquad i = 0, 1, ..., m$$

where $w_{ia}(Y, \dot{Y}) = \dot{W}_{ia}(Y_i)\dot{Y}_i - \partial T_{ia}(Y_i, \dot{Y}_i)/\partial Y_i + D_{ia}\dot{Y} + 0.5\partial(Y^T \overline{\pi}_a Y)/\partial Y_i +$ $+ \pi_{ia}(Y)Y - G_i(m_i\, g) \in R^{6\times 1}$. By putting together all $6(\mathrm{m}+1)$ equations it is obtained

$$W_a(Y)\ddot{Y} + w_a(Y, \dot{Y}) = F \quad (6.33)$$

where $W_a(Y) = diag(W_{0a}(Y_0)\; W_{1a}(Y_1)\; ...\; W_{ma}(Y_m)) \in R^{6(m+1)\times 6(m+1)}$, $W_a(Y) = W_a^T(Y)$, det $W_a(Y) \neq 0$, $w_a(Y, \dot{Y}) = (w_{0a}^T(Y, \dot{Y}) ... w_{ma}^T(Y, \dot{Y}))^T \in R^{6(m+1)\times 1}$. From $6(m+1)$ equations (6.33) only rank K equations are independent. Equation (6.33) can be presented in such a way that the descriptions of connections and manipulated object motion are separated

$$W_{ca}(Y_c)\ddot{Y}_c + w_{ca}(Y, \dot{Y}) = F_{ca}$$

$$W_{0a}(Y_0)\ddot{Y}_0 + w_{0a}(Y, \dot{Y}) = 0 \tag{6.33a}$$

where index c designates the quantities connected to the contact points, and index 0 designates the quantities connected to the manipulated object. Here $Y_{ca} = (Y_{1a}^T\, Y_{2a}^T \ldots Y_{ma}^T)^T \in R^{6m\times 1}$, $Y_{0a} \in R^{6\times 1}$, $F_{ca} = (F_{1a}^T\, F_{2a}^T \ldots F_{ma}^T)^T \in R^{6m\times 1}$, $F_{0a} = 0 \in R^{6\times 1}$, $W_{ca}(Y_{ca}) = diag(W_{1a}(Y_1) \ldots W_{ma}(Y_m)) \in R^{6m\times 6m}$, $W_{ca}(Y_c) = W_{ca}^T(Y_c)$, $\det W_{ca}(Y_c) \neq 0$, $w_{ca}(Y, \dot{Y}) = (w_{1a}^T(Y, \dot{Y}) \ldots w_{ma}^T(Y, \dot{Y}))^T \in R^{6m\times 1}$, where the expanded position vector of the contact position in the $6m$-dimensional space is denoted by Y_{ca} and the expanded vector of the contact forces acting at the contact point in that space is denoted by F_{ca}. It should be noted that at the MC of the manipulated object no contact force acts directly, hence $F_{0a} = 0$. Equations (6.33) and (6.33a) represent the final form of equations of elastic system behavior under the action of external forces F_{ca} performing the general motion about the unloaded state 0, while the unloaded state 0 is also performing general motion.

General motion of elastic system motion is described by (6.33), i.e. $6(\mathrm{m}+1)$ relations are defined, rank K of which is independent. This means, that for the unique definition of elastic system position during motion it is necessary to prescribe $6(m+1) - \text{rank } K$ absolute generalized coordinates and their derivatives.

2.4 Model of Manipulator Dynamics

Model of the motion of a nonelastic manipulator with six DOFs with rigid (uncompliant) joints and with a force at the gripper in the internal coordinates space is given in [21–26]

$$H_i(q_i)\ddot{q}_i + h_i(q_i, \dot{q}_i) = \tau_i + J_i^T f_{ci},\ i = 1, \ldots., m \tag{6.34}$$

where $H_i(q_i) \in R^{6\times 6}$ is a positive definite inertia matrix of the $i-$th manipulator, $h_i(q_i, \dot{q}_i) \in R^{6\times 1}$ is the vector that encompasses the influences of gravity, centrifugal, and Coriolis' effects, $\tau_i \in R^{6\times 1}$ is the vector of the drives in joints, $J_i \in R^{6\times 6}$ is the transformation matrix, by which the velocity vector is mapped from the internal coordinates into the manipulator tip vector, and $f_{ci} = -F_i \in R^{6\times 1}$ is the contact force at the manipulator gripper. For *m* manipulators the model is generalized in the form

$$H(q)\ddot{q}+h(q,\dot{q})=\tau+J^T f_c \tag{6.35}$$

where $H(q)=diag(H_1(q_1),\ldots, H_m(q_m)) \in R^{6m\times 6m}$, $h(q,\dot{q})=(h_1^T(q_1,\dot{q}_1),\ldots, h_m^T(q_m,\dot{q}_m))^T \in R^{6m\times 1}$, $\tau=(\tau_1^T,\ldots,\tau_m^T)^T \in R^{6m\times 1}$, $J^T=diag(J_1^T,\ldots,J_m^T)\in R^{6m\times 6m}$, $f_c=(f_{c1}^T,\ldots,f_{cm}^T)^T=(-F_1^T,\ldots,-F_m^T)^T \in R^{6m\times 1}$, $q=(q_1^T,\ldots,q_m^T)^T \in R^{6m\times 1}$, $\dot{q}=(\dot{q}_1^T,\ldots,\dot{q}_m^T)^T \in R^{6m\times 1}$. By the vector equation (6.35) *6m* connections are determined. If the kinematic connection of the internal and absolute coordinates of contact points is designated by $Y_i=\Theta_i(q_i)\in R^{6\times 1}$, $i=1,\ldots,m$, then the relation between their velocities and accelerations is given by $\dot{Y}_i=\partial\Theta_i(q_i)/\partial q_i\cdot\dot{q}_i=J_i(q_i)\,\dot{q}_i\in R^{6\times 1}$, $\ddot{Y}_i=\dot{J}_i(q_i)\dot{q}_i+J_i(q_i)\ddot{q}_i\in R^{6\times 1}$, *i=1,…,m,* or in the united form

$$\begin{aligned}&Y_c=\Theta(q)\in R^{6m\times 1},\ \dot{Y}_c=J(q)\dot{q}\in R^{6m\times 1},\\&\ddot{Y}_c=\dot{J}(q)\dot{q}+J(q)\ddot{q}\in R^{6m\times 1}\end{aligned} \tag{6.36}$$

where $Y_c=(Y_1^T,\ldots,Y_m^T)^T$.

2.5 Model of Cooperative Manipulation

By the equations (6.33), (6.35), and (6.36) the cooperative work of m rigid manipulators with six DOFs, handling a rigid object, onto the general motion of which in a $3D$ space no constraints are imposed, and whereby the connections between object and manipulators are elastic, has been defined. The number of model inputs is $6m$, while the number of independent state quantities (positions and velocities) is $2\cdot 6(m+1)$, wherein $2\cdot rank\,K$ are defined by the elastic system and $2\cdot(6(m+1)-rank\,K)$ are defined by the leading manipulator dynamics.

The state quantities can be chosen in different ways. For instance, possible choices are: the vector of internal coordinates $q\in R^{6m\times 1}$ and the position vector of the manipulated object MC $Y_0\in R^{6\times 1}$, or the position vector of the MC connections and of the manipulated object $Y\in R^{6(m+1)\times 1}$, or the vector of internal coordinates of leading manipulator $q_v\in R^{(6(m+1)-rankK)\times 1}$ and position vector of the resting connections MC, and of the manipulated object MC $Y_a\in R^{rank K\times 1}$. Let the state vector be introduced as

$z=(q^T Y_0^T)^T \in R^{6(m+1)\times 1}$. As $F_{ca}=-f_c$, by substituting (6.36) and (6.33a) into (6.35) it is obtained

$$N(q)\ddot{q}+n(q,\dot{q},Y_0,\dot{Y}_0)=\tau,$$

$$W(Y_0)\ddot{Y}_0+w(q,\dot{q},Y_0,\dot{Y}_0)=0 \qquad (6.37)$$

where $N(q)=H(q)+J^T(q)W_{ca}(\Theta(q))J(q)\in R^{6m\times 1}$, $n(q,\dot{q},Y_0,\dot{Y}_0)=h(q,\dot{q})+$ $+J^T(q)W_{ca}(\Theta(q))\dot{J}(q)\dot{q}+J^T(q)w_{ca}(\Theta(q),J(q)\dot{q},Y_0,\dot{Y}_0)\in R^{6m\times 1}$, $W(Y_0)=W_0(Y_0)$, $w(q,\dot{q},Y_0,\dot{Y}_0)=w_{0a}(\Theta(q),J(q)\dot{q},Y_0,\dot{Y}_0)\in R^{6\times 1}$. If vector $z=(q^T Y_0^T)^T$ is introduced into (6.37), the general form of the cooperative work model is obtained

$$\Phi(z)\ddot{z}+\phi(z,\dot{z})=\delta_\tau, \qquad \det\ \Phi(z)\neq 0 \qquad (6.38)$$

where $\Phi(z)=diag(N(q),W(Y_0))\in R^{6(m+1)\times 6(m+1)}$, $\phi(z,\dot{z})=col(n(q,\dot{q},Y_0,\dot{Y}_0),$ $w(q,\dot{q},Y_0,\dot{Y}_0))\in R^{6(m+1)\times 1}$, $\delta_\tau=(\tau^T\ 0\ldots 0)^T\in R^{6(m+1)\times 1}$, $\tau\in R^{6m\times 1}$ whereby the essential characteristics of cooperative work follows that the number of possible (physical) inputs is smaller than the number of model state variables (system order).

2.6 Example

The presented modeling procedure is illustrated by an example of cooperative work of three manipulators each with three DOFs, handling an object with mass center CM_0 while contacts are being established in the mass centers of the connections CM_1, CM_2, and CM_3 (in Fig. 6.3, initial position of cooperative system is given). In the example, the state vector is $(Y_0^T q^T)^T = (Y_0^x Y_0^y Y_0^\varphi q_1^1 q_1^2 q_1^3 \ldots q_3^1 q_3^2 q_3^3)^T \in R^{12\times 1}$ and the model has the following form (according to (6.37), (6.36), and (6.33a)):

$$\ddot{q}=N^{-1}(q)\,(\tau-n(q,\dot{q},Y_0,\dot{Y}_0))\ \left[rad/s^2\right]\in R^{9\times 9},$$
$$\ddot{Y}_0=-W^{-1}w(q,\dot{q},Y_0,\dot{Y}_0)\ \left[m/s^2\right]\in R^{3\times 1}$$
$$F_c=W_c(\dot{J}(q)\dot{q}+J(q)\ddot{q})+w_c(q,\dot{q},Y_0,\dot{Y}_0)\ [N]\in R^{9\times 1}$$

where $N(q)=H(q)+J^T(q)W_cJ(q)\in R^{9\times 9}$, $n(q,\dot{q},Y_0,\dot{Y}_0)=h(q,\dot{q})+J^T(q)W_c\dot{J}(q)\dot{q}+$ $+J^T(q)w_c(q,\dot{q},Y_0,\dot{Y}_0)\in R^{9\times 1}$. According to (6.35) it is $H(q)=diag\,(H_1(q_1),$ $H_2(q_2),\ H_3(q_3))\in R^{9\times 9}$ and in the chosen example the coefficients are:

$H_i^{11}(q_i^2) = I_i^1 + (a_i^1)^2 m_i^1 + I_i^2 + ((L_i^2)^2 + (a_i^2)^2 + 2L_i^1 a_i^2 \cos q_i^2)\, m_i^2 + I_i^3 +$ $+((L_i^1)^2 + (L_i^2)^2 + 2L_i^1 L_i^2 \cos q_i^2) m_i^3$, $H_i^{12}(q_i^2) = l_i^2 + ((a_i^2)^2 + L_i^1 a_i^2 \cos q_i^2) m_i^2 +$ $+ I_i^3 + ((L_i^2)^2 + L_i^1 L_i^2 \cos q_i^2) m_i^3$, $H_i^{22} = I_i^2 + (a_i^2)^2 m_i^2 + I_i^3 + (L_i^2)^2 m_i^3$, $H_i^{13} = H_i^{23} = H_i^{33} = I_i^3$, where $L_i^j = 1\,[m]$, $i = 1,2,3$, $j = 1,2$ are the lengths of manipulators links, $a_i^j = 0.25[m]$, $i = 1,2,3$, $j = 1,2$ distances of the links MCs from the rotation points, $m_i = (3,3,2)\,[kg]$, $i = 1,2,3$ and $I_i = (0.25,$ $0.25, 0.01)\,[kgm^2]$, $i = 1,2,3$ are the vectors of masses and moments of inertia of the manipulator links. Vector components $h\,(q,\dot{q}) = col\,(h_1(q_1,\dot{q}_1),$ $h_2(q_2,\dot{q}_2),\ h_3(q_3,\dot{q}_3)) \in R^{9\times1}$ are given by $h_i^1(q_i,\dot{q}_i) = -L_i^1(a_i^2 m_i^2 +$ $+ L_i^2 m_i^3)\dot{q}_i^2(\dot{q}_i^2 + 2\dot{q}_i^1)\sin q_i^2$, $h_i^2(q_i,\dot{q}_i) = -L_i^1(a_i^2 m_i^2 + L_i^2 m_i^3)\dot{q}_i^1(\dot{q}_i^1 + 2\dot{q}_i^2)\times$ $\times \sin q_i^2$ and $h_i^3(q_i,\dot{q}_i) = 0$, $i = 1,2,3$. Matrices W_c and W are constant diagonal matrices $W_c = diag(m_1, m_1, A_1, m_2, m_2\, A_2, m_3, m_3, A_3) = diag(1, 1,$ $0.005, 1, 1, 0.005, 1, 1, 0.005) \in R^{9\times9}$ and $W = W_0 = diag\,(m_0, m_0, A_0) =$ $= diag\,(5, 5, 0.1) \in R^{3\times3}$, where $m_0 = 5$ [kg], $m_i = 1$[kg], and $A_0 = 0.1[kgm]$, $A_i = 0.005[kgm]$, i = 1,2,3 are the mass and moments of inertia of the body CM_0 and of the connections CM_1, CM_2, and CM_3.

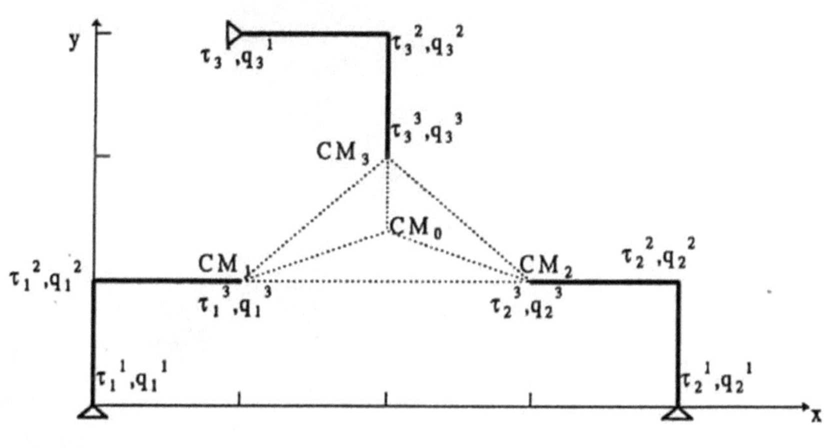

Figure 6.3. Initial position of cooperative system.

Vector $w_a = w_a(q,\dot{q},Y_0,\dot{Y}_0) \in R^{12\times1}$ from (6.33) is formed by the vectors $w = w(q,\dot{q},Y_0,\dot{Y}_0) \in R^{3\times1}$ and $w_c = w_c(q,\dot{q},Y_0,\dot{Y}_0) \in R^{9\times1}$ as $w_a = col(w, w_c) =$

$=D_a\dot{Y}+\pi_a Y+0.5\partial(Y^T\bar{\pi}_a Y)/\partial Y\in R^{12\times 1}$. The damping submatrices are $D_{ij}=diag(\mathcal{G}_{ij}\,D_{ij}^{\delta}\,\mathcal{G}_{ij},d_{ij}^{\varphi})\in R^{3\times 3}$, $ij=01,02,03,12,13,23$, where $r_{ia}=(x_{ia}\,y_{ia})^T$, i=1, 2, 3, $\mathcal{G}_{ij}=(r_{ia}-r_{ja})\cdot(r_{ia}-r_{ja})\,/\left\|r_{ia}-r_{ja}\right\|^2\in R^{2\times 2}$, whereby the norm of vector is $\left\|\rho_{ija}\right\|=\left\|r_{ia}-r_{ja}\right\|^2=((x_{ia}-x_{ja})^2+(y_{ia}-y_{ja})^2)^{0.5}$. The constant damping matrix D_{ij} is formed by the damping of linear motion, given in the diagonal matrix $D_{ij}^{\delta}=\text{diag}\ (d_{ij}^{x},\,d_{ij}^{y})\in R^{2\times 2}$, and the damping of rotational motion d_{ij}^{φ}. The adopted values are $d_{ij}^{x}=d_{ij}^{y}=20\ [N/(m/s)]$ and $d_{ij}^{\varphi}=15[Nm/(rad/s)]$, ij=01,02,03,12,13,23. Submatrices of the stiffness matrix are defined by $\pi_{ij}=(c_{ij}^{x}(1-(\left\|\rho_{ij0}\right\|/\left\|r_{ia}-r_{ja}\right\|))^2,c_{ij}^{y}(1-(\left\|\rho_{ij0}\right\|/$ $/\left\|r_{ia}-r_{ja}\right\|))^2,c_{ij}^{\varphi})\in R^{3\times 3}$, where the coefficients $c_{ij}^{x}=c_{ij}^{y}=1225[N/m]$ and $c_{ij}^{\varphi}=25\ [Nm/rad\ ]$, ij=01, 02, 03, 12, 13, 23. The derivatives of the stiffness submatrices are determined by (sign * is used to replace x and y, and index "*ind*" to replace any particular index):

$$\frac{\partial\pi_{ij}}{\partial *_{ind}}=\begin{cases}0,\ *_{ind}\neq x_{ia},\,x_{ja},\,y_{ia},\,y_{ja}\\[2ex] 2\,(*_{ind}-*_{ja})\,(1.-\dfrac{\left\|\rho_{ij0}\right\|}{\left\|r_{ia}-r_{ja}\right\|})\,\dfrac{\left\|\rho_{ij0}\right\|}{\left\|r_{ia}-r_{ja}\right\|^3}\,diag\,(c_{ij}^{x},c_{ij}^{y},0),\ *=x,y,\ ind=ia\\[2ex] -2(*_{ia}-*_{ind})\,(1.-\dfrac{\left\|\rho_{ij0}\right\|}{\left\|r_{ia}-r_{ja}\right\|})\,\dfrac{\left\|\rho_{ij0}\right\|}{\left\|r_{ia}-r_{ja}\right\|^3}\,diag\,(c_{ij}^{x},c_{ij}^{y},0),\ *=x,y,\ ind=ja\end{cases}$$

where $\left\|\rho_{ij0}\right\|=((x_{i0}-x_{j0})^2+(y_{i0}-y_{j0})^2)^{0.5}$ and ij= 01, 02, 03, 12, 13, 23.

Connection (6.36) is defined by

$$Y_i=\Theta_i(q_i)=\begin{bmatrix}L_i^1\cos\,q_i^1+L_i^2\cos\,(q_i^1+q_i^2)+Xb_i\\ L_i^1\sin\,q_i^1+L_i^2\sin\,(q_i^1+q_i^2)+Yb_i\\ q_i^1+q_i^2+q_i^3\end{bmatrix}\in R^{3\times 1},\quad i=1,\,2,\,3$$

where $(Xb_i, Yb_i) = (0,0), (4,0)$ and $(1,3)\,[m,\, m]$ are the coordinates of the $i-$th manipulator base, so that the corresponding Jacobian matrices and their derivatives are

$$J_i(q_i)=\begin{bmatrix} -L_i^1 \sin q_i^1 - L_i^2 \sin(q_i^1+q_i^2) & -L_i^2 \sin(q_i^1+q_i^2) & 0 \\ L_i^1 \cos q_i^1 + L_i^2 \cos(q_i^1+q_i^2) & L_i^2 \cos(q_i^1+q_i^2) & 0 \\ 1 & 1 & 1 \end{bmatrix} \in R^{3\times 3}, \quad i=1,2,3$$

$$\dot{J}_i(q_i)=\begin{bmatrix} -L_i^1 \dot{q}_i^1 \cos q_i^1 - L_i^2 (\dot{q}_i^1+\dot{q}_i^2) \cos(q_i^1+q_i^2) & -L_i^2 (\dot{q}_i^1+\dot{q}_i^2) \cos(q_i^1+q_i^2) & 0 \\ -L_i^1 \dot{q}_i^1 \sin q_i^1 - L_i^2 (\dot{q}_i^1+\dot{q}_i^2) \sin(q_i^1+q_i^2) & -L_i^2 (\dot{q}_i^1+\dot{q}_i^2) \sin(q_i^1+q_i^2) & 0 \\ 0 & 0 & 0 \end{bmatrix} \in R^{3\times 3}$$

The expanded Jacobian matrix and its derivative are given by:

$$J(q) = diag(J_1(q_1), J_2(q_1), J_3(q_1)) \in R^{9\times 9}$$
$$\dot{J}(q) = diag(\dot{J}_1(q_1), \dot{J}_2(q_1), \dot{J}_3(q_1)) \in R^{9\times 9}.$$

Simulation results for $\tau_i^j = 0,\ i,\ j = 1,\ldots,3$, except $\tau_1^1 = 50[Nm]$, $\tau_2^1 = -50\ [Nm]$, and for the initial state defined by Fig. 6.3, are given in Fig. 6.4. The components of the realized contact forces at the contact points CM_1 and CM_2 along the axis x, $F_1^x = -50[N]$ and $F_2^x = 50[N]$ can serve as check values.

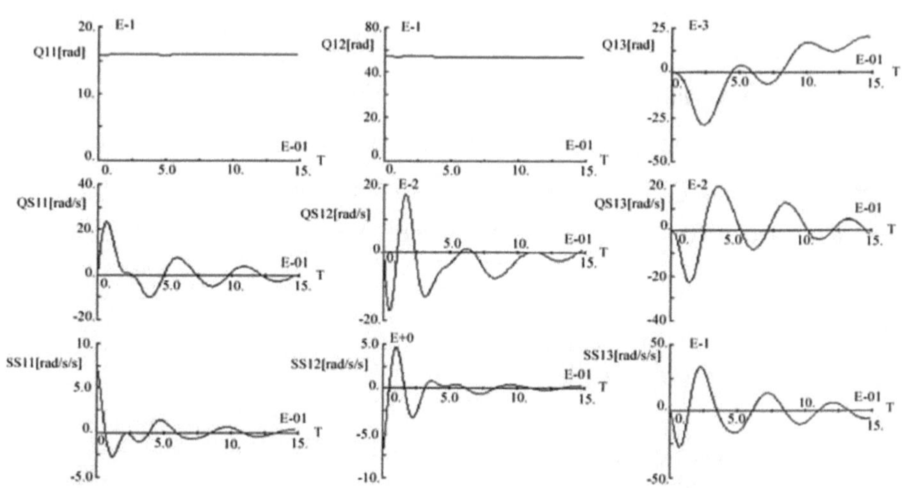

Figure 6.4.(a) Simulation results.

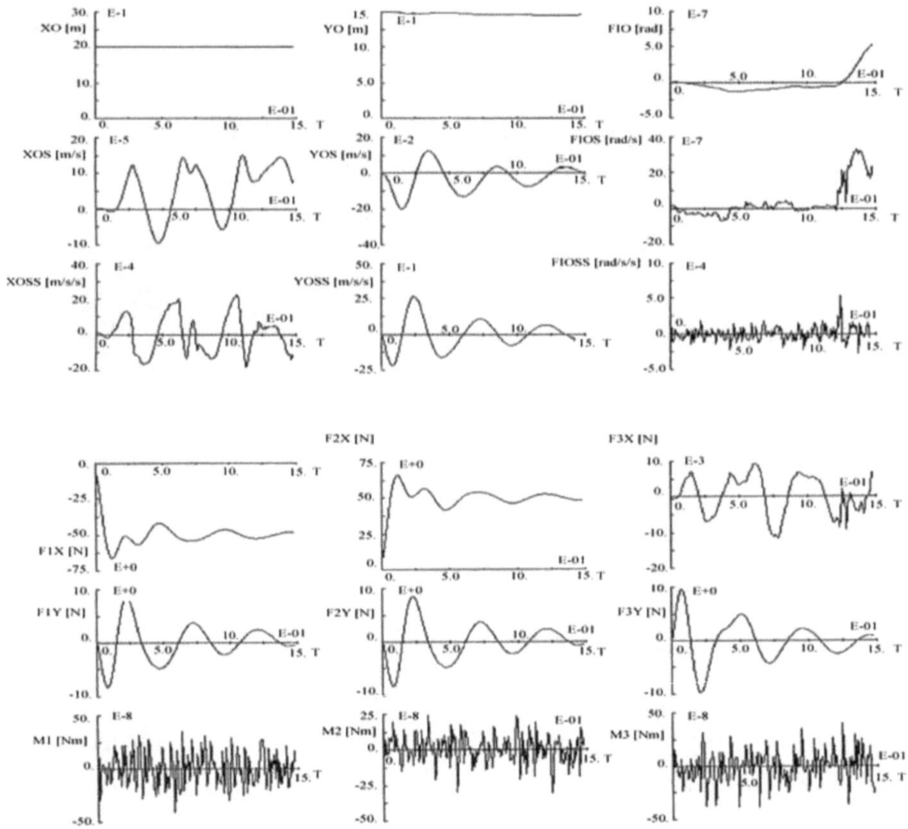

Figure 6.4.(b) Simulation results (in all the diagrams the independent variable is time in seconds).

The dependent variables are designated as: $Q13$ is the symbol for internal coordinate q_1^3, while $QS13$ and $SS13$ are symbols for its first and second derivative $\dot{q}_1^3$ and $\ddot{q}_1^3$. Position symbols CM of the manipulated object are $X0$, $Y0$, and $FI0$, while FiX, FiY, and Mi, $i = 1, 2, 3$ are those of the contact forces.

3. CONCLUSION

The procedure of generating the complete mathematical model of dynamics of nonredundant multi-arm cooperative robots, based on Lagrange's equations, is given. The obtained model is presented in several forms. The problem of indefiniteness with respect to force distribution was fully resolved, i.e. the reason for the emergence of the problem of indefiniteness with respect to force and kinematic indefiniteness in cooperative work is given. It was shown that this problem does not exist physically, and the procedure for solving the indefiniteness with respect to force in accordance with physical phenomena is presented. According to this, the problem of kinematic indefiniteness in cooperative work, was also redefined. The system of active spatial mechanisms with six DOFs elastically interconnected with the object (dynamic environment), is modeled. General motion of a system of elastically connected rigid bodies with six DOFs is modeled, starting from the assumption that the absolute coordinates of the body MC during motion, as well as distances between the MCs, are known before the deformation of the elastic system began. The modeling procedure and the model were illustrated by an appropriate example.

REFERENCES

1. S. Hayati, "Hybrid Position/Force Control of Multi-Arm Cooperating Robots", Proc. of IEEE Conf. on Robotics and Automation, San Francisco, pp. 82-89, Apr. 1986.
2. Y. Nakamura, K. Nagai, and T. Yoshikawa, "Mechanics of Coordinative Manipulation by Multiple Robotic Mechanisms", Proc. of IEEE Conf. on Robotics and Automation, Raleigh, pp. 991-998, Mar.-Apr. 1987.
3. K. Kosuge, J. Ishikawa, K. Furuta, and M. Sakai, "Control of Single-Master Multi-Slave Manipulator System Using Vim," Proc. of IEEE Conf. on Robotics and Automation, Cincinnati, Ohio, pp. 1172-1177, May 1990.
4. A.B. Cole, J.E. Hauser, and S.S. Sastry, "Kinematics and Control of Multifingered Hands with Rolling Contact," IEEE Trans. of Automatic Control, Vol. 34, pp. 398-404, Apr. 1989.
5. Y.F. Zheng, and J.Y.S. Luh, "Optimal Load Distribution for Two Industrial Robots Handling a Single Object," Proc. of IEEE Conf. on Robotics and Automation, Philadelphia, pp. 344-349, Apr. 1988.
6. A. Ramadorai, T. Tarn, and A. Bejczy, "Task Definition, Decoupling and Redundancy Resolution by Nonlinear Feedback in Multi-Robot Object Handling", Proc. of IEEE Conf. on Robotics and Automation, Nice, France, pp. 467-474, May 1992.
7. K. Jankowski, H. Elmaragfy, and W. Elmaragfy, "Dynamic Coordination of Multiple Robot Arms with Flexible Joints", Int. J. Robotic Research, Vol. 12, pp. 505-528, Dec. 1993.
8. M. Djurovic, and M. Vukobratovic, "A Contribution to Dynamic Modeling of Cooperative Manipulation", Mechanism and Machine Theory, Vol. 25, No. 4, pp. 407-416, 1990.
9. I. Walker, S. Marcus, and R. Freeman, "Distribution of Dynamic Loads for Multiple Cooperating Robot Manipulators", Journal of Robotic Systems, Vol. 6, No. 1, pp. 35-47, 1989.
10. I. Prokofyev, Theory of Constructions (in Russian), Nauka, Moscow, 1963.
11. J. Argyris, and S. Kelsey, Energy Theorems and Structural Analysis, Butterworths, London, 1971.

12. S. Rao, The Finite Element Method in Engineering, Pergamon Press, Oxford, 1982.
13. R. Gallagher, Finite Element Analysis (in Russian). Englewood Cliffs, New Jersey: Prentice-Hall, 1975.
14. Z. Luo, and M. Ito, "Control Design of Robot for Compliant Manipulation on Dynamic Environments", IEEE Trans. of Robotics and Automation, Vol. 9, pp. 286-296, June 1993.
15. Z. Luo, K. Ito, and M. Ito, "Multiple Robot Manipulators' Compliant Manipulation on Dynamical Environments", Proceedings of IROS, pp. 1927-1934, 1993.
16. Z. Luo, Y. Uematsu, K. Ito, A. Kato, and M. Ito, "On Cooperative Manipulation of Dynamic Object", Proceedings of IROS, pp. 1423-1431, 1994.
17. P. Akella, and M. Cutkosky, "Contact Transition Control with Semiactive Soft Fingertips", IEEE Trans. of Robotics and Automation, Vol. 11, pp. 859-867, Dec. 1995.
18. O. Al-Jarrah, Y. Zheng, and K. Yi, "Trajectory Planning for Two Manipulators to Deform Flexible Materials Using Compliant Motion", Proc. IEEE Int. Conf. on Robotics and Automation, Nagoya, Japan, pp. 1517-1522, 1995.
19. D. Sun, X. Shi, and Y. Liu, "Modeling and Cooperation of Two-Arm Robotic System Manipulating a Deformable Object", Proc. of IEEE Conf. on Robotics and Automation, Minneapolis, MN, pp. 2346-2351, Apr. 1996.
20. V. Novozilov, Osnovi nelinjejnoj teorij uprugosti (in Russian), Ogiz, Leningrad, Gosudarstvjenoje Izdeljstvo Tehniko-teoreticeskoj Literaturi, 1948.
21. M. Vukobratovic, Applied Dynamics of Manipulation Robots, Springer-Verlag, 1989.
22. M. Vukobratovic, and D. Stokic, Control of Manipulation Robots, Springer-Verlag, 1989.
23. M. Vukobratovic, and Y. Ekalo, "Unified Approach to Control Laws Synthesis for Robotic Manipulators in Contact with Dynamic Environment", Tutorial S5: Force and Contact Control in Robotic Systems, IEEE Int. Conf. on Robotics and Automation, Atlanta, pp. 213-229, 1993.
24. M. Vukobratovic, and Y. Ekalo, "New Approach to Control of Robotic Manipulators Interacting with Dynamic Environment", Robotica, Vol. 14, pp. 31-39, 1996.
25. Y. Ekalo, and M. Vukobratovic, "Adaptive Stabilization of Motion and Forces in Contact Tasks for Robotic Manipulators with Non-Stationary Dynamics", Int. Journal on Robotics and Automation, Vol. 9, pp. 91-98, 1994.
26. Y. Ekalo, and M. Vukobratovic, "Quality of Stabilization of Robots Interacting with Dynamic Environment", Journal of Intelligent and Robotic Systems, Vol. 14, pp. 155-178, 1995.

Chapter 7

NEW TRENDS IN CONTACT TASKS

In principle, one can distinguish two basic subclasses of contact tasks. One subclass essentially covers force tasks, whose very nature requires an end-effector to establish physical contact with the environment and exert a process-specific force. In general, these tasks require both the position of the end-effector and the interaction force to be controlled simultaneously. A typical example of such tasks is machining processes such as grinding, deburring, polishing, bending, etc. In these tasks, force is an inherent part of the process and plays a decisive role in its execution (e.g., metal cutting or plastic deformation). In order to prevent overloading or damaging of the tool during the operation, force must be controlled in accordance with some definite task requirements.

The prime emphasis of the tasks from the other subclass lies on the end-effector motion that has to be realized close to the constrained surfaces (compliant motion). A typical representative of such tasks is the part mating process. The problem of controlling the robot during these tasks is, in principle, the problem of accurate positioning. However, due to imperfections inherent in the process, sensing, and control system, these tasks are inevitably accompanied by the occurrence of contact with the constrained surfaces, which results in the appearance of reaction forces. The measurement of interaction force provides useful information for error detection and appropriate modification of the prescribed robot motion.

The future will certainly hold more tasks for which interaction with the environment is of fundamental importance. Recent medical robot applications in surgery, e.g., spine surgery, neurosurgical and microsurgical operations [1], and hip implant operations [2], could also be considered essentially as contact tasks. Comprehensive research programs in automated construction [3], agriculture, and the food industry [4], focus on the robotization of several representative contact tasks, such as underground excavation, meat deboning, etc.

In the following text we will present in brief several examples of contact tasks that are currently occupying the attention of researchers and professionals, and whose common specific feature is the contact of the system (structure) with its dynamic environment.

1. ROAD VEHICLE

We consider the road vehicle as a dynamic system that has to overpower the reaction forces that tend to stop the vehicle or change its direction. The inertial and elasto-damping properties of the tire pneumatics, viscous shock absorbers, and spring dampers have a predominant influence on the intensity of the so-called contact load forces. The intensities of these forces determine also the values of the coefficients of tire rolling and sliding friction and the amplitudes of their longitudinal and lateral forces. In this consideration it is assumed that the vehicle is a rigid body which relies on the support through the so-called suspension system (Fig. 7.1).

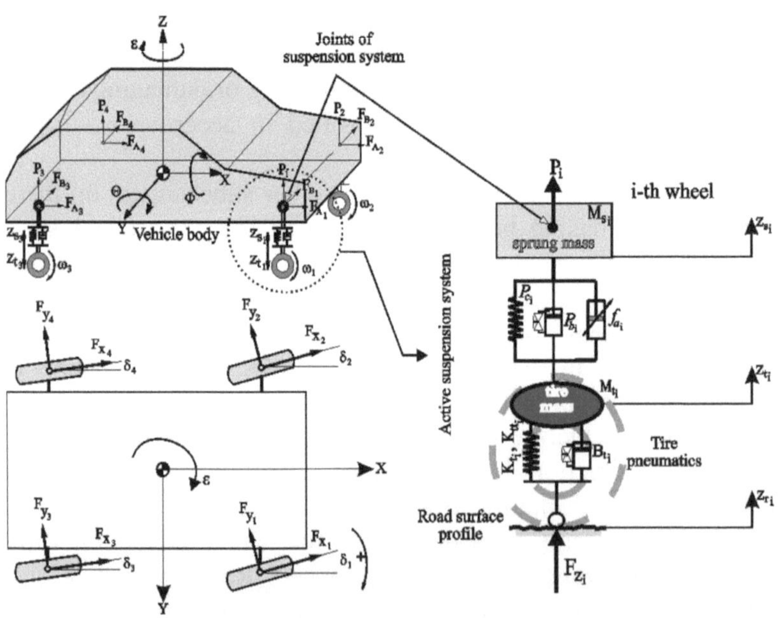

Figure 7.1. Road vehicle as the rigid body supported by 4-wheel suspension system

It can be conditionally assumed that the vehicle suspension system and the ground make in a wider sense the vehicle's "dynamic environment". Such functional partition of the road vehicle into vehicle body and its "environment" can be effectively used for control purposes [5, 6]. Assuming that the vehicle moves on a solid ground, dynamic characteristics of the suspension system have a dominant influence on the dynamic behavior of the vehicle with respect to the support. The road surface itself can be practically considered as a kinematic constraint on the vehicle motion.

It should be emphasized that the dynamic structure is in contact with its dynamic environment, which can be either of a passive or active character. In the first case, as was already mentioned, passive dynamic environment represents the suspension together with elastodynamic and damping parameters, whereas active dynamic environment involves automatic changes of the environment parameters in accordance with the changeable intensity of the external dynamic load. In both cases, one should also take into account the dynamics of tire pneumatics as part of the vehicle dynamic environment.

In such complex problems of dynamics and control (semiautomatic and especially automatic) with respect to position and force, it is undoubtedly of essential importance to have as faithful as possible representation of the dynamic situation in which the complex mechanical system is in a dynamic environment, in the form of a direct contact of the structure (construction) with the suspension and, via the dynamics of tire pneumatics, with a ground of different character. At that, when a direct dynamic environment of the road vehicle is involved, it is especially important to adopt active suspension, which undoubtedly represents a contemporary tendency in manufacturing road vehicles of high dynamic performance.

2. RAILWAY VEHICLE

A dynamic picture of the motion of two-axis or four-axis railway cars in dynamic contact with the rails is of essential importance in constructing an active suspension system, and automatic maintenance of the lateral force of the contact between the wheels and rail, responsible for possible derailing, which can have catastrophic consequences, especially in the present time of enhanced speeds of railway vehicles.

In the case of such a task we can observe the dynamic contact between the car and the rail, which has its elastodynamic characteristics in contrast to the conventional solution where the rail is considered as a kinematic constraint and only its kinematic inaccuracy (i.e. the deviation from an ideally manufactured rail) is taken into account.

Figures 7.2 and 7.3 illustrate the example of dynamic complexity of the task described above, whereas Figure 7.4 shows a hypothetical realization of the active system motion of a railway vehicle. Even so, a complex task can have its natural extension in the form of the dynamic contact of the railway vehicle using active pantograph, whose purpose is to ensure approximately constant forces of the contact of the pantograph with the catenary, which makes the problem of maintaining desirable performance of motion of the railway vehicle more complex in respect of comfort and safety of passengers.

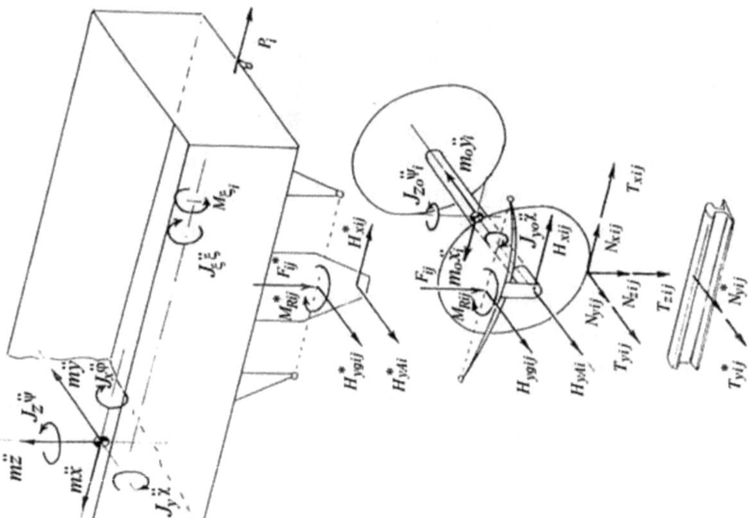

Figure 7.3. Connection between rail and

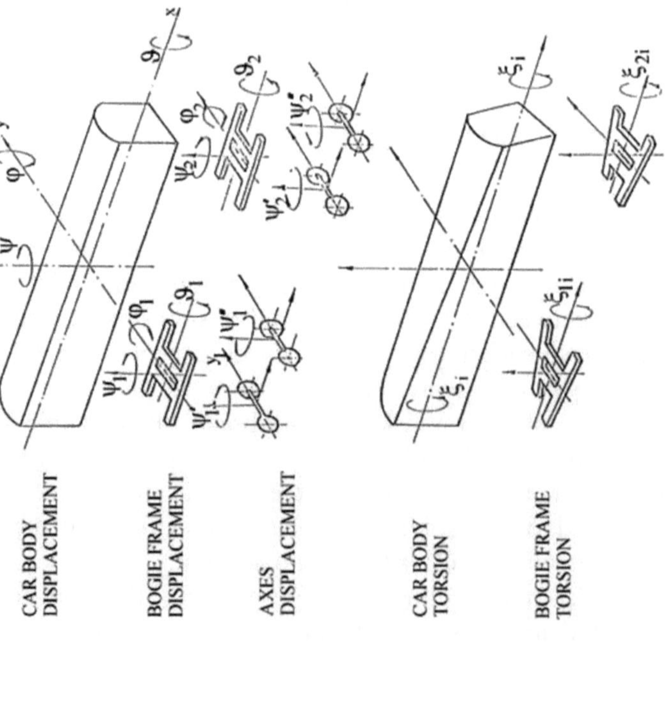

Figure 7.2. Degrees of freedom of railway vehicle

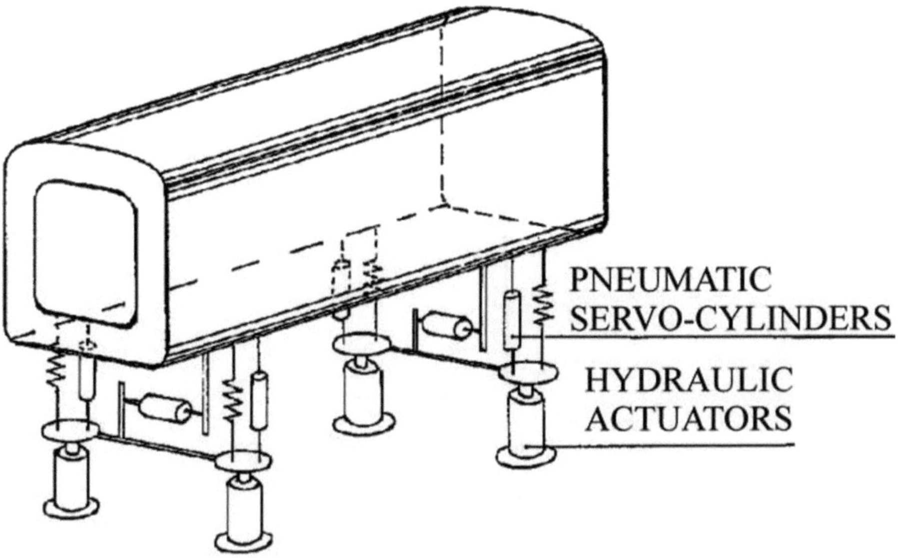

Figure 7.4. Schematic presentation of active control of railway vehicle

3. HUMAN AND HUMANOID DYNAMICS

Biped locomotion has been at the focus of research for decades. Theoretical studies from various aspects have been accompanied by many simulations and realizations: from the simplest cases of planar mechanisms to humanoid robots, the most complex locomotion mechanisms constructed to date. During walking, two different situations arise in sequence: the statically stable double-support phase in which the mechanism is supported on both feet simultaneously, and statically unstable single-support phase, when only one foot of the mechanism is in contact with the ground. Thus, the locomotion mechanism changes its structure during the single walking cycle from an open to a closed kinematic chain.

All of the biped mechanism joints are powered and directly controllable except for the "joint" formed by the contact of the foot with the ground. This contact is essential for the walk realization because this is the only point at which the mechanism interacts with the environment, and the mechanism's position relative to the environment depends on the regularity of its behavior. The foot behavior cannot be controlled directly, but in an indirect way, by ensuring appropriate dynamics of the mechanism above the foot. Thus, the overall indicator of the mechanism behavior is the ground reaction force: its intensity, direction, and particularly its acting point. This point, termed Zero-Moment Point (ZMP) [7–9], is shown in Fig. 7.5.

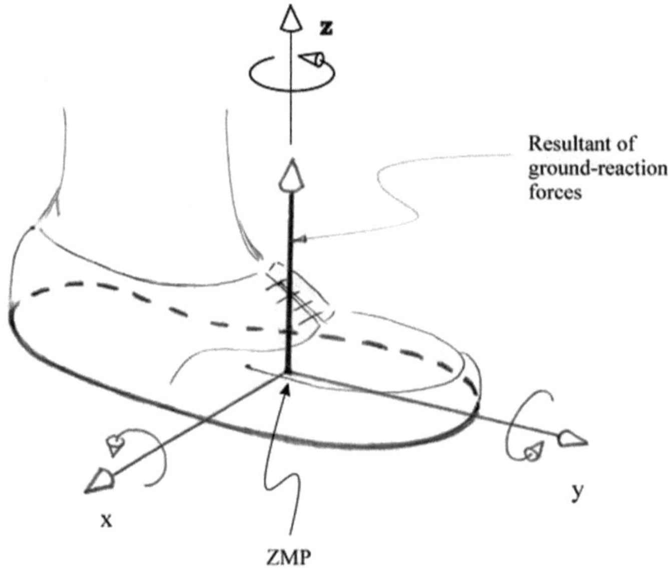

Figure 7.5. Zero-moment point (ZMP)

It is obvious that human walk represents a specific contact task and a relatively high degree of faithfulness is required to mimic it in the artificial walk of humanoid robots. A number of humanoid robots whose mechanical complexity exceeds forty degrees of freedom have already been constructed. The point that is of interest in relation to humanoid robotics, and which we address in this short communication, is concerned with the phenomena of dynamic contacts. For an effective control of a biped gait, it is necessary to control the contact forces between the feet and the ground. To do this efficiently, the deformation of the two bodies being in contact has to be taken into account. Deformation may appear on the foot side and on the ground side (in this article we will use the more appropriate term *support* instead of ground). So, this topic relates the contact zone and its properties on both sides.

In robot dynamics, the foot is generally considered rigid. However, in the contact zone this assumption may appear inappropriate. With the human foot (barefoot), the sole is not rigid – the skin and the subcutaneous and connective tissues are deformable. If a shoe or a robot foot is considered, then it is reasonable to expect again a **flexible sole** – some rubber layer introduced to act as a shock absorber. Such effects, elastodynamic contact, have been elaborated in industrial robotics but the experience with humanoids is limited. For a general simulator, it is necessary to introduce this elastic layer and it should feature six-component deformation: lateral displacements along three axes and angular displacements about these axes (Fig. 7.6). It is generally assumed that deformation is a dynamic effect rather than just a spring and damper. Hence, we model the sole flexibility by means of appropriate impedance. One should note that the sole deformation influences the robot absolute position (lateral and angular, relative to the external frame), and also robot dynamics and stability.

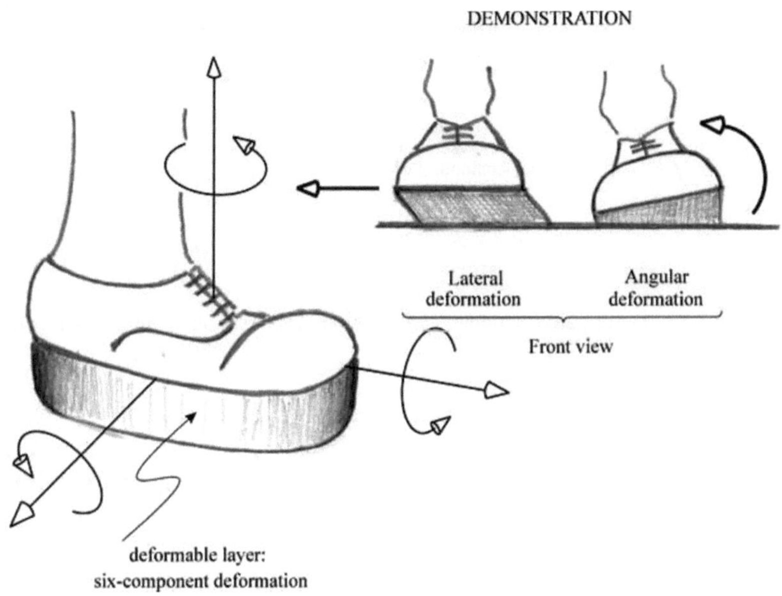

Figure 7.6. Deformable sole: deformation has six components; two components are demonstrated

Now, let us look at the surface on which the human or the robot walks. In a general case, this **support features elastic properties**: it can deform along the three axes and about them (six-component deviation – see Fig. 7.7). Let us justify the work on this effect. One can imagine training of special forces whereby the trainees have to move on some "soft" support, such as a sponge or a water mattress. Simulation of these actions involves elastodynamics of the support. In another example, a sportsman on a trampoline might be simulated. The deformation of the support is modeled in the same way as it was done on the foot side of the contact – introducing appropriate impedance. This deformation influences all the positions and the entire dynamics "above" it.

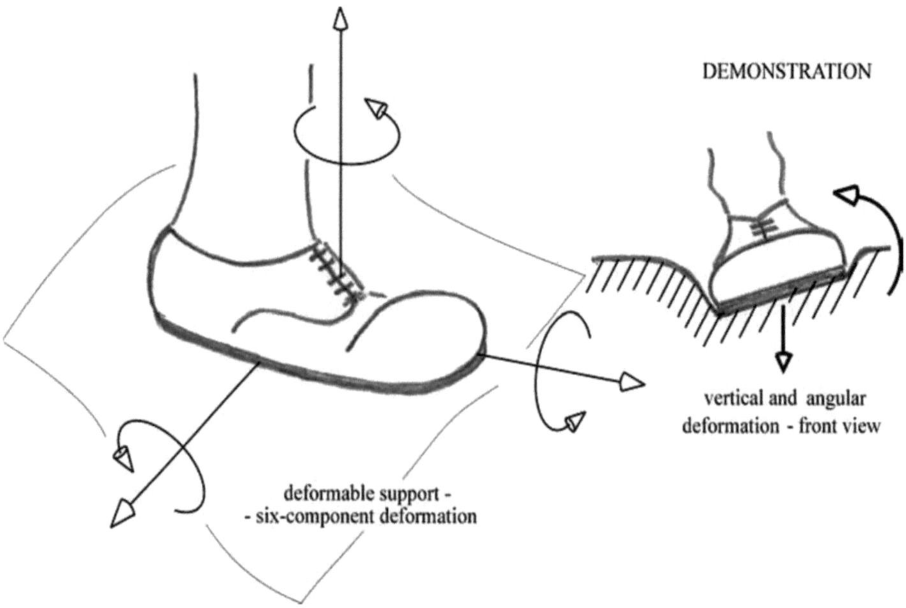

Figure 7.7. Deformable support: deformation has six components; two components are demonstrated

The introduction of elasticity in the contact zone allows the exact treatment of the **impact** that occurs when the foot touches the support. Although the elasticity eliminates the impact shock, there are still effects that need analysis (such as transient response, etc.).

Here, a natural question arises: why the two deformations (one on the foot side and the other on the support side) are considered separately? There are several reasons for that. One is that the two deformations come out from different layers and may have different characteristics. The next regards the gait stability. If the walker loses its stability (dynamic balance), the foot (together with its shock-absorbing layer) overturns about the edge, being between the two deformable layers (Fig. 7.8). In order to analyze the stability of gait, ZMP position is considered. It refers to a mobile frame (laterally shifted and inclined).

The next focus is on the support. We have already stated that the term ground in this context should be replaced by *support*. The term ground usually means something immobile (perhaps deformable, but immobile). However, a general approach requires the option of walking on a **mobile support**. Moreover, such support is not considered as a pure nonstationary constraint but rather as a dynamic system that interacts with the walker (or a jumper, ...). Thus, a mobile platform that has its own dynamics is introduced [10]. The platform should have up to six DOFs.

Justification for this generalization comes out from real needs. Training of special forces and police includes different actions performed on a mobile platform. Another example might be simulation of walking on a ship or a boat.

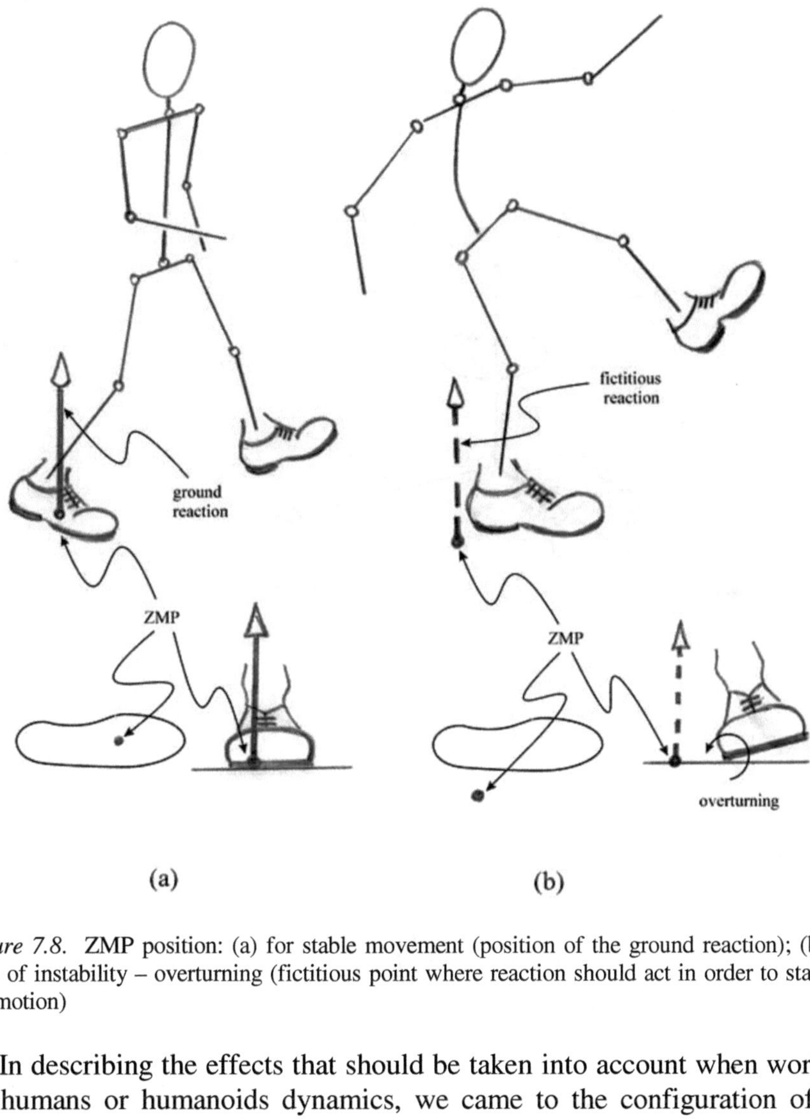

Figure 7.8. ZMP position: (a) for stable movement (position of the ground reaction); (b) the case of instability – overturning (fictitious point where reaction should act in order to stabilize the motion)

In describing the effects that should be taken into account when working on humans or humanoids dynamics, we came to the configuration of the system shown in Fig. 7.9. It is clear that a complex contact system is at stake. With the so augmented mechanical system, the movement is still synthesized and stability is controlled by using ZMP concept. One should note that ZMP conditions now refer to a mobile frame (laterally shifted and inclined).

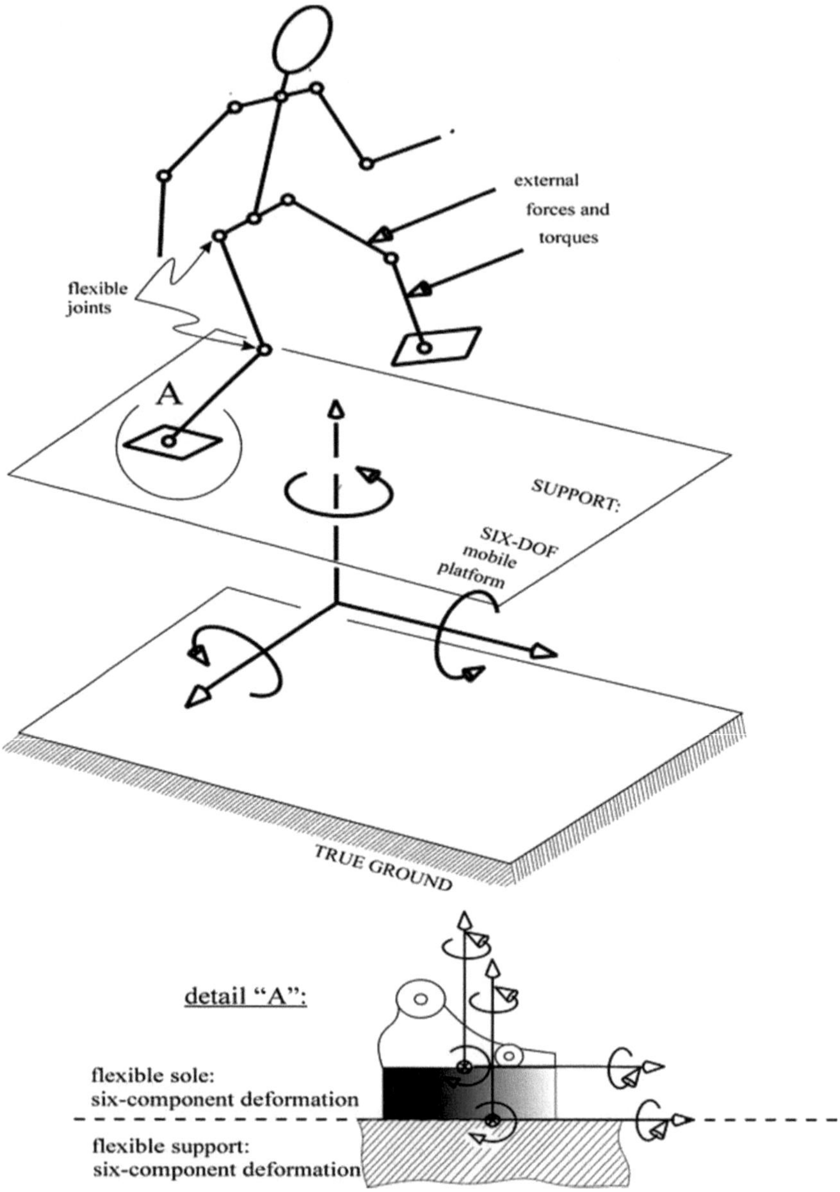

Figure 7.9. General configuration of the system

Another question deserves an answer. What kinds of motions are to be modeled? One could note that we mainly used the terms *walking* or *gait*. This came out from the fact that walking was the most interesting problem and the topic that has attracted major attention for a long time. However, many other movements are of interest, such as balancing on one foot or two feet, running, jumping, etc. Skiing is another interesting motion that might be modeled and simulated. It is specific because of sliding which is the key effect.

4. OPERATOR–VIBRATIONAL PLATFORM

The discussion of contact tasks can also be extended to the problem of contact of man (operator) and a platform subjected to programmed perturbations of different character (Fig. 7.10).

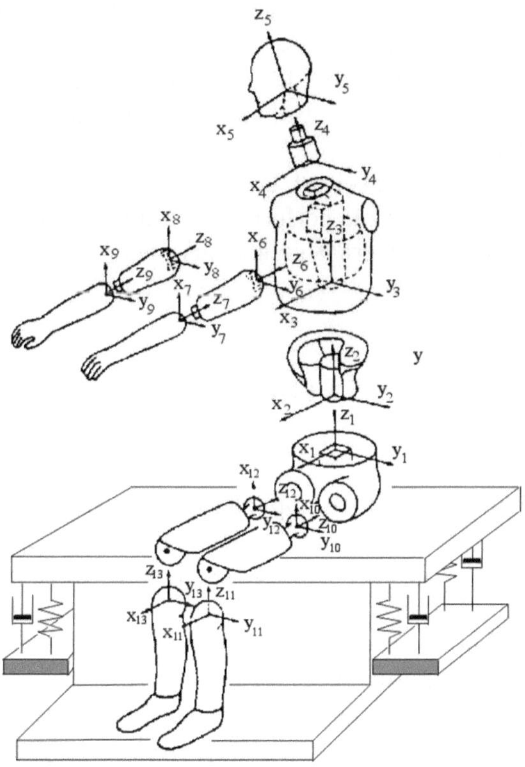

Figure 7.10. Mechanical model of the operator–platform interaction

Harmful vibrations of such a platform can be analyzed by the appropriate description of the dynamics of the interacting systems – mechanical model, representing a sufficiently faithful imitation of the real operator and the vibrational platform, specific vehicle equipped with the operator's seat, represented also by the corresponding mechanical model.

Of course, such a scheme can also be supplemented by a manipulation subsystem (Fig. 7.10) that will perform certain manipulation tests in the course of vibrations produced by a defined spectrum of vibrational regimes.

Here we can distinguish various working regimes of the interactive control complex. Let us formulate only one of them.

The subsystems involved in the interaction are spatial powered mechanism of the simulated operator in contact with the dynamic platform system, including also dynamic module of the seat. This interactive dynamic complex can be considered in two versions – passive and active.

In the passive version, the dynamic modules have no possibility of automatically changing their characteristic dynamic parameters, while in the active version these interactive subsystems possess the possibility of changing their pertinent dynamic parameters.

Such tasks, if considered correctly, have to be solved by applying simultaneous dynamic position–force control [11] (see Appendix 2) with the test of practical stability included [12].

The active version of the conceived contact task assumes also the possibility of maintaining the desirable performance of the overall system in the presence of the changes of external conditions under which its stable operation should be ensured.

5. ROBOTIC SYSTEMS FOR THERAPY AND MASSAGE

The advantages of robots where their application in the field of medicine is concerned are the more precise and more reliable actions, independent of physician's nerve stress and subjective estimations, as well as the possibility of numerous repetitions of the procedure.

At present, medical robots can be classified with regard to their invasion features. Invasive robots perform resections, biopsies, and punctures. Minimal invasive robots move into cavities or vessels, performing surgical operations or diagnostics. Non-invasive robots perform manipulations on soft tissues, for example, in the techniques of minimal therapy and massage.

Manipulations on soft tissues are realized by mechanical action of hands or tools on soft tissues, causing deformations of some parts of these tissues (mechanoreceptors, biological vital points, vessel walls) and changing the state of human on the whole, or local parts of his body and organs. The soft tissues are the elastic biological ones: skin, fat layers, muscles, fascias.

From the point of view of mechanics the massage movements are spatial displacements of a masseur's hands or tool, involving a tremendous variety of movements. There are straight line, arc, spiral, and screw trajectories. The trajectories differ because of a great range of parameters of patient's surface relief. Depending on the magnitude and direction of soft tissue deformation, the massage techniques are called stroking, squeezing, kneading, ribbing, pinching, or tension [13].

The idea of controling apparatus for massage by means of a computer was suggested for the first time at the Second Medical Robotics Workshop [13].

Robotic massage represents an extremely specific contact task of the robot in interaction with its dynamic environment. The specificity of this interaction stems from the fact that the environment is in fact the human body (i.e. soft tissues) and from a great variety of manual therapy and massage techniques, the requirement of high safety and reliability, etc.

This system has many uncertainties. Firstly, the robot tool trajectory is not assigned and programmed in advance, nor are the forces of interaction of the robot tool with soft tissues. The pattern of the movement is defined only approximately. Secondly, the position of the patient relative to the manipulator is not defined precisely; it may change during the procedure. Thirdly, mechanical parameters of soft tissues (inertia, viscosity, elasticity) vary for different individuals and body parts, and may change during the procedure. Fourthly, effective values of biomechanical and biomedical parameters and their relation to mechanical action parameters are not defined completely. The essence of these uncertainties arises from the presence of two human objects (physician and patient) in the system. Thus, fuzzy logic seems to be most appropriate to control this kind of system, and the model may be neurofuzzy [13].

In Fig. 7.11 is presented a photograph of the first realization of the experiment of robotic massage carried out at the Moscow State Industrial University, Deptartment of Automatics, Informatics and Control Systems [14].

On the basis of robotic massage as a specific task of contact type, where the robot interacts with its environment, it can be concluded that it is necessary to form a dynamic model of the environment involved (in this case the human body) [11,15].

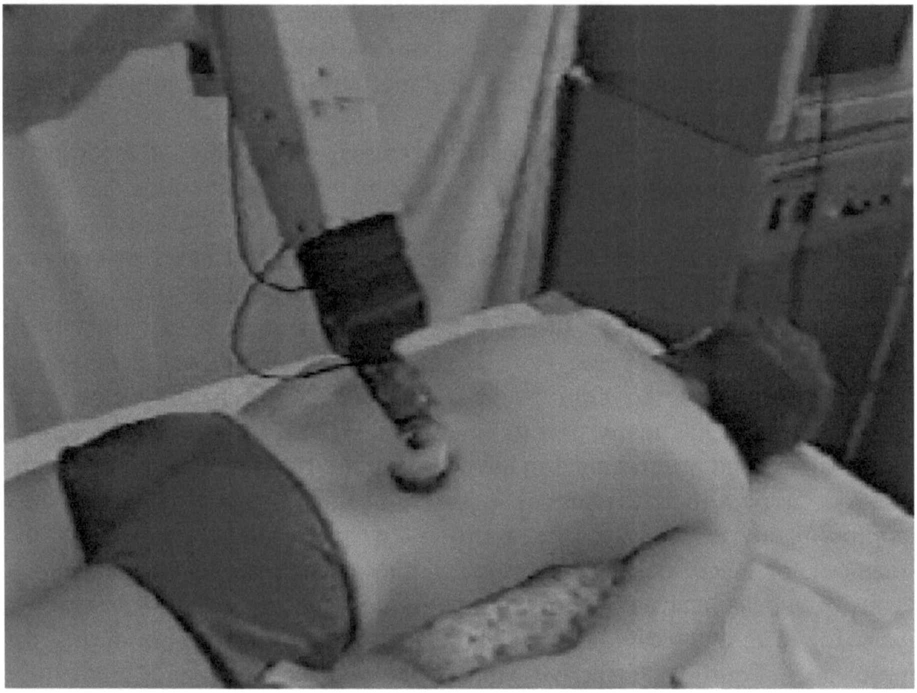

Figure 7.11. Robotic massage experiment

On the basis of the developed position–force dynamic control laws [11] it is necessary to apply simultaneous control with respect to both position and force in order to ensure stable position (motion) for a stable interaction force [12].

When speaking of this unconventional contact task it is necessary to point out some difficulties that stand on the way of the integrated dynamic position–force control of robots interacting with dynamic environments.

This is primarily related to the uncertainties of dynamic parameters of the environment, especially to the biodynamic parameters of its equivalent mass, stiffness, and damping. Because of this, in studying stability with respect to the position and contact force we have to choose the preset limits of the estimated range of the above parameters for which we carry out tests of practical stability of the overall interactive system, robot plus dynamic environment. Here, one more remark should be added – the tests of practical stability thus carried out are of a conservative character because of a series of linearization operations (approximations) of expressions in the process of execution of the test of sufficient stability conditions [12].

6. CONCLUDING REMARKS

In concluding this brief discussion of the previous and potential future accounts of contact tasks involving essentially dynamic systems in the interaction with dynamic environment, we would like to point out that there exist a wide range of tasks in technical practice that require appropriate treatment of the objects in direct contact with the environment of very different dynamic nature: building structures and constructions under the dynamic loads of air pressure distribution, seismic acceleration of the soil in contact with object (constructions), missiles guided in the air and water, and contact between structures, such as arch dams (i.e. their foundations) in contact with the environment (soil), etc.

Despite the fact that essential issues in modern technical and technological practice are related to control, a detailed and comprehensive dynamic analysis, for which nowadays there are practically no technological limitations (barriers), is an essential prerequisite for the synthesis of reliable dynamic control laws, especially those that are concerned with contact tasks.

In connection with contact tasks, of which several nonstandard examples have been presented in this chapter, it is interesting to note that the majority of them can be related to the active systems that are essentially capable of automatically adjusting dynamic parameters of the environment in order to maintain the necessary performance of the system (object, structure) under the changeable (sometimes extremal) conditions of their functioning.

After this observation it is necessary to say something about dynamic environment, which is not of a strictly defined (invariable) character. Namely, the problem remains of determining the equivalent mass of the dynamic environment with which the object (structure, functional system) interacts. It is certainly true that part of the equivalent mass (portion of the mass that influences the interaction) of the dynamic environment will depend on the magnitude of elastodynamic deflection due to the different intensities of contact force, i.e. the technological conditions of the contact task.

Apart from a satisfactory identification of dynamic parameters of the equivalent mass (inertia), equivalent damping, and stiffness of the environment, it is of essential importance to have the possibility of their automatic adaptation, to preserve the desired performance of the systems in contact.

Based on the mentioned examples of real or hypothetical active systems, we can speak of very specific robotized constructions, structures, or systems.

Thus, we can already speak about unconventional robotized (active) structures, such as suspension bridges, buildings, cooling towers, pilotless/ driverless vehicles, etc. It is quite realistic to expect in the near future widespread realization of active systems that will lead to an essential improvement of their dynamic performance, particularly under conditions when the capabilities of traditional (passive) systems are exhausted.

In conclusion, we take the liberty of trying to somehow resolve the purposefully posed dilemma on whether the present development of active systems is a temporary euphoria or a real necessity. The authors are inclined to the latter option bearing in mind the selective needs for active systems, which are, as a rule, based on the dynamics analysis of the contacting systems, their interactive dynamic models, and the synthesis of dynamic laws of simultaneous position–force control, as well as the test of their practical stability.

Hence the authors thought it best to produce a relatively slim volume that would make the reader not feel helpless when confronted with the challenge of formulating and solving numerous new tasks, basically of contact character, encountered in modern technical practice.

REFERENCES

1. S. Ku, S.E. Salcudean, "Design and Control of a Teleoperated Microgripper for Microsurgery", Proceedings of IEEE International Conference on Robotics and Automation, Minneapolis, pp. 889-894, 1996.
2. D.W. Heikkinen, B. Musits, "Sankyo Robotics and ROBODOC – A Story of International Cooperation", Proceedings of the 24th International Symposium on Industrial Robotics (ISIR), Tokyo, Japan, Session C-4, 1994.
3. G.H. Watson, R.L. Tucker, J.K. Walters (editors), Automation and Robotics in Construction X, Proceedings of the 10th International Symposium on Automation and Robotics in Construction (ISARC), Elsevier, Amsterdam, 1993.
4. P.T. Clarke, S.J. Key, P. Tang, "Development of Robot Technology for Meat Deboning", Proceedings of International Symposium on Robots of the Australian Robot Association and the International Federation of Robotics, Sydney, Australia, pp. 427-439, 1988.
5. A. Rodic, M. Vukobratovic, Dynamics, Integrated Control and Stability of Automated Road Vehicles, Ibidem-Verlag, Stuttgart, 2002.
6. A. Rodic, M. Vukobratovic, "Design of an Integrated Active Control System of Road Vehicles", Intern. Journal of Computer Application in Technology, special issue on Active Structures, Vol. 13, Nos. 1/2, 2000.
7 .M. Vukobratovic, D. Juricic, "Contribution to the Synthesis of Biped Gait", IEEE Trans. on Automatic Control, Vol. 16, No 1, 1969.
8. D. Juricic, M. Vukobratovic, "Mathematical Modeling of a Bipedal Walking System", Proc. of the ASME winter annual meeting, (72-WA/BHF-13, New York), Nov. 26-130, 1972.
9. M. Vukobratovic, "How to Control Artificial Anthropomorphic Systems", IEEE Trans. on Systems, Man and Cybernetics, Vol. SMC-3, Sept. 1973.
10. M. Sorli, M. Vukobratovic, C. Ferraresi, M. Kolarski, B. Borovac, " Mechanics of Turin Parallel Robot", Mechanism and Machine Theory, Vol. 32, No 1, pp. 51-77, 1997.

11 M. Vukobratovic, Yu. Ekalo, "New Approach to Control of Robotic Manipulators Interacting with Dynamic Environment", Robotica, Vol. 14, pp. 31-39, 1996.

12 D. Stokic, M. Vukobratovic, "Practical Stabilization of Robots Interacting with Dynamic Environment", Int. Journal of Robotics and Automation, Vol. 13, Issue 4, 1998.

13 V. Golovin, "Robot for Massage", Proceedings of JARP 2nd Workshop on Medical Robotics, Heidelberg, 1997.

14 V. Golovin, A. Samorukov, The New Method of Apparatus Massage, protected by a Russian patent No 2145833, 1998.

15 A. De Luca, C. Manes, "Hybrid Force/Position Control for Robot in Contact with Dynamic Environment", Proc. Robot Control SYROCO, Vienna, 1991.

APPENDIX 1

HYBRID POSITION/FORCE CONTROL

This approach is based on a theory of compliant force and position control formalized by Mason [1] and is concerned with a large class of tasks involving partially constrained robotic motion. In this approach, depending on the specific mechanical and geometrical characteristics of the given contact problem, one can distinguish two sets of constraints upon the robot's motion and the contact forces. A set of constraints which is a natural consequence of the task configuration, i.e. of the nature of the desired contact between the end-effector held by the robot and constraining surface, is called *natural constraints.* A suitable frame in which natural constraints are prescribed is referred to as the *task* or *constraint frame* [2]. In order to specify the task that should be realized by the robot in the compliant frame, the so-called *artificial constraints* have to be introduced. Practically, these constraints partition the possible DOFs of the motion into those which must be position-controlled and those which should be force-controlled in order to perform the given task. It is reasonable to define an artificial constraint with respect to force when there is a natural constraint on the gripper motion in this direction (DOF), and vice versa. For the sake of simplicity, a diagonal Boolean matrix S, $\overline{S} = I - S$, called the compliance selection matrix [3], has been introduced. In accordance with the prescribed artificial constraints the i-th diagonal element of this matrix has the value 1 if the i-th DOF with respect to the task frame is to be force controlled, and the value 0, if it is position controlled.* Once the contact task is specified, the next step is to

* It is assumed that the task frame coincides with the Cartesian frame.

select the appropriate control algorithms. The relevant methods are discussed below [4, 5].

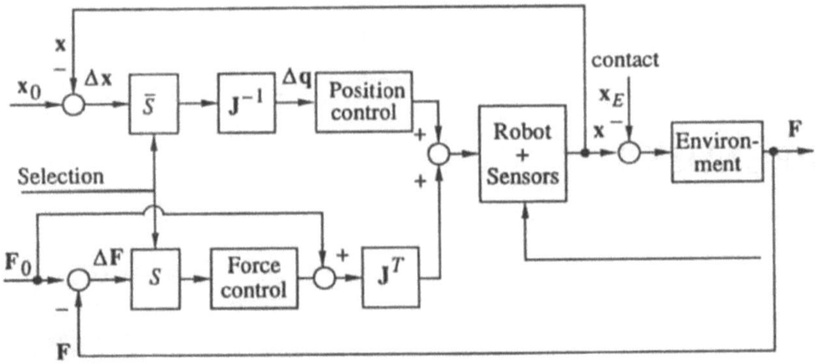

Figure A.1.1. Explicit hybrid position/force control (Raibert and Craig 1981)

The most important method within this group is certainly the algorithm proposed by Raibert and Craig [3]. Figure A.1.1 represents the control scheme which illustrates the main idea of this method. The control consists of two parallel feedback loops, the upper one for the position, and the lower one for the force feedback loop. Each of these loops use separate sensor systems. The positional loop utilizes the information obtained from the positional sensors at the robot joints, and the force loop is based on force sensing data. Separate control laws are adopted for each loop. The central idea of this hybrid control method is to apply two independent control loops assigned to each DOF in the task frame. Both control loops cooperate synchronously, to control each of the manipulator joints.

This concept, at first glance, appears to be ideally established to solve the hybrid position/force control problems. However, with a deeper insight into the method, some essential difficulties and problems related to this concept appear. The first problem is related to the opposite requirements of the hybrid control concept concerning position and force control subtasks. In fact, it is required that the position control is highly stiff in order to keep positioning errors in the selected directions as small as possible, while the force control requires relatively low stiffness of the robot (corresponding to the desired force) in the force-controlled direction with respect to the task frame, in order to ensure that the end-effector behaves compliantly with the environment. As already explained above, the explicit hybrid control tries to solve this problem by control decoupling into two independent parts which are position and force controlled (Fig. A.1.1). It would mean that the position control part does not interfere with the force-control loop. However, that is not the case. Since each robot joint contributes to the control of both position and force, couplings in the manipulator mechanical structure (implied in the Jacobian matrix) cause that the control input to the actuator corresponding to the force loop (e.g., force-controlled directions) produces additional forces in position-controlled directions in the task frame, and vice versa. Due to this coupling between position- and force-controlled DOFs by setting the position errors in the force-controlled directions to zero, the position feedback gains in all directions are actually reduced, causing the entire system to become more sensitive to perturbations. As a consequence, the robot's performance is not unique with this scheme for all configurations and for all force-commanded directions. Moreover, one can find certain configurations for which, depending on selected force and position directions, the system becomes unstable. Since only the robot's position and the selection matrix influence the instability, this phenomenon is referred to as *kinematic instability* [6].

In order to overcome these crucial problems of the above algorithm, Zhang [7] has recently proposed to introduce an addtional selection of input forces (Fig. A.1.1). In other words, the input torques from positional- and force-controlled parts are decoupled in the task-frame before they are applied to the joints. Fisher and Mujtaba [8] proposed a generalized projection matrix to decouple position- and force-controlled direction. Thus the system becomes stable, but similar to the original control scheme, the eigenvalue of the system changes not only with variation of the robots configuration, but also with the given task, i.e. the selection matrix. This causes the robot performance to be strongly dependent on the configuration and selection of controlled directions.

The second problem refers to some dynamic stability issues in force control, such as high-gain effect of force sensor feedback (caused by high environment stiffness), unmodeled high frequency dynamic effects (due to arm and sensor elasticity), contact with stiff environment, noncollocated sensing and control, and is named the *dynamic stability problem* [9].

To overcome dynamic problems of the hybrid position/force control several researchers have pursued the idea of including the robot dynamic model into the control law. The resolved acceleration control, originally formulated for the position control [10], belongs to the group of dynamic position control algorithms [5]. Shin and Lee [11] have extended this approach to the hybrid position/force control. The joint space implementation of the proposed control scheme is sketched in Fig. A.1.2.

In this scheme, the driving torques compensate for the gravitational, centrifugal, and Coriolis' effects, and feedback gains are adjusted according to the changes in the inertial matrix. An acceleration feedforward term is also included to compensate for the changes along nominal motion in position directions. Finally, the control inputs are computed by

$$\tau = \hat{\Lambda}\bar{S}\ddot{x}^* + \hat{\mu}(x,\dot{x}) + \hat{p}(x) + Sf^* \tag{A.1.1}$$

where $\ddot{x}^*$ is the commanded equivalent acceleration

$$\ddot{x}^* = \ddot{x}_0 + K_v(\dot{x}_0 - \dot{x}) + K_p(x_0 - x), \tag{A.1.2}$$

and f^* is the command vector from the force control parts whose form depends on the applied control law. To minimize the force error, it is convenient to introduce a PI force regulator of the form

$$f^* = K_{f_p}(F_0 - F) + K_{f_r}\int(F_0 - F)\,dt \tag{A.1.3}$$

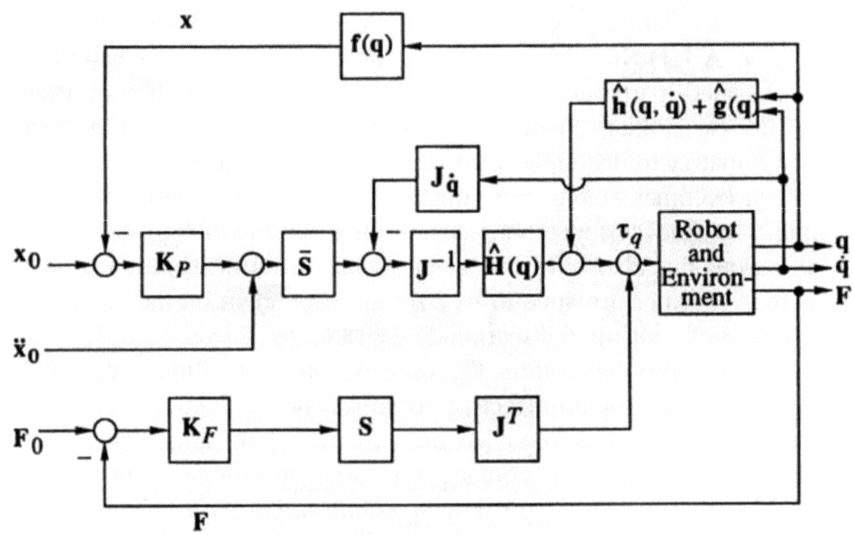

Figure A.1.2. Resolved acceleration – motion force control.

Khatib [12] has introduced an "active damping" term into the force control part to avoid bouncing and minimize force overshoots during transition (impact effects):

$$\tau_f = Sf^* - \hat{\Lambda}SK_{vf}\dot{x} \tag{A.1.4}$$

where K_{vf} is a diagonal Cartesian damping matrix. Other quantities will be explaned in the next footnote, where the superscripts ^ represent the estimates of these quantities.

Further modifications of the control scheme have been proposed by Bona and Indri [13]. To compensate for the coupling between the force and position control loops, as well as for the disturbance on the position controller due to the reaction force, the authors have modified the position control law according to:

$$\tau_p = \hat{\Lambda}\overline{S}[\ddot{x}^* - \hat{\Lambda}^{-1}(Sf^* - F)] + \hat{\mu}(x,\dot{x}) + \hat{p}(x) \qquad \text{(A.1.5)}$$

If the dynamic modeling used for computation of the control law is correct, the above control law provides complete decoupling between the position and force control in the task frame, i.e. ensures the following closed-loop behavior:

$$\ddot{x} = \overline{S}\ddot{x}^* + S\hat{\Lambda}^{-1}Sf^* - S\hat{\Lambda}^{-1}F - SK_{vf}\dot{x} \qquad \text{(A.1.6)}$$

The next group of algorithms consider more complicated constraints on the robot motion which are described by a set of rigid hypersurfaces in the space of end-effector Cartesian coordinates [14] or in the joint coordinates space [15]** The system model is described with a typical set of linearly-

** When the manipulator interacts with the environment it is very convenient to describe its dynamics in the space where manipulation task is described, rather than in joint coordinate space (also termed configuration space). The end-effector position and orientation with respect to a reference coordinate system can be described by a 6-dimensional vector *x*. For the sake of simplicity we consider a conventional industrial robot with 6 DOF.

The dynamic models of the actuators and mechanical part of the robot are related by joint torques (loads) (2.85, 2.86). If we substitute τ_a (vector of generalized joint – axes driving torques) from

$$n_i^2 I_{m_i}\ddot{q}_{m_i} + n_i^2 b_{m_i} + \tau_{a_i} = n_i \tau_{m_i}$$

with q_{m_i} (output angle of motor shaft after reducer), n_i (gear ratio), I_{m_i} (inertia of the actuator), b_{m_i} (motor viscous friction coefficient), τ_{m_i} (control input to the i-th actuator, i.e. motor torque), into

$$H(q)\ddot{q} + h(q,\dot{q}) + g(q) = \tau_q + J(q)^T F\,,$$

we get the entire model of the robotic mechanism in joint coordinate space

$$\overline{H}(q)\ddot{q} + B_m\dot{q} + h(q,\dot{q}) + g(q) = \tau_q + J(q)^T F$$

where $\overline{H} = H + I_m = H + diag(n_i^2 I_{m_i}),\ B_b = diag(n_i^2 b_{m_i})$.

The above dynamic model can be transformed into an equivalent form that is more convenient for analysis of a robot controller for contact tasks. The relationships between the internal (joint) and external (end-effector) robot coordinates is defined by nonlinear, continuous, and differentiable functions $x = f(q)$. By differentiating this equation with implicit second-order differential – algebraic equations (mechanical differential – algebraic equations). This model is used to compute the control law comprising the following functions: linearization and decoupling of the system dynamics, as well as decomposition of the control problem into position- and force-controlled directions.

To improve reliability of the concept, the dynamic hybrid control is extended into the case of unknown environments, consisting of hypersurfaces [16]. The improved control schemes involve on-line identification algorithms based on force and position measurements, and adaptive control mechanisms. However, as already mentioned, the adaptive constrained motion control is theoretically attractive, but still impractical in reality. Hence, the hybrid control algorithms become even more complex and difficult to implement in real time with usual computational and sensing resources presently available with robotic manipulators.

The hybrid position/force task specification has also been subject of considerable investigations. Lipkin and Duffy [17] demonstrated that the Mason's position/force decomposition approach based on "geometrical orthogonality" is in fact erroneous. The resulting planning for hybrid control are invariant neither with respect to a translation of origin, nor with respect to the change of unit length. The authors proposed a more general and mathematically consistent invariant hybrid task formulation based on screw algebra. In this approach the complementarity between motion (modeled by a twist) and force (represented by a wrench) is expressed via the *reciprocity relationship,* which is independent of coordinate frame, scaling, or units. Two fundamental relations between twist and wrench, referred to as *freedom* and *constraints equations,* have been introduced to test task compatibility with the model of environment. These relations correspond to analytical

---

respect to time we obtain: $\dot{x} = \dfrac{\partial f(q)}{\partial q^T}\dot{q} = J(q)\dot{q}$. Using the velocity mapping relation and the derived relationships between accelerations: $\ddot{x} = J\ddot{q} + \dot{J}\dot{q}$, as well as between forces specified in the operational and corresponding joint torques: $\tau_{qF} = J(q)^T F$, we can transform the robot dynamic model from the joint into the end-effector coordinate system:

$$\Lambda(x) = J(q)^{-T}\overline{H}(q)J(q)^{-1}, \quad B(x) = J(q)^{-T}B_m J(q)^{-1},$$

$$\mu(x,\dot{x}) = J(q)^{-T}h(q,\dot{q}) - \Lambda(x)\dot{J}(q,\dot{q})\dot{q}, \quad p(x) = J(q)^{-T}g(q), \quad \tau = J(q)^{-T}\tau_q .$$

Analogues to the joint space quantities, $\Lambda(x)$ is the operational space inertia matrix, $\mu(x,\dot{x})$ is the vector of Coriolis' and centrifugal forces, $p(x)$ is the vector of gravity terms, and τ is the applied input control force in the operational configuration space.

expressions for natural and artificial constraints in non-invariant hybrid approach. For every constrained motion task two screws subspaces can be derived that correspond to artificial constraints. These subspaces represent the sets of screws about which twists and wrenches can be controlled.

In some simple tasks and reference frames, both conventional and reciprocity-based decomposition show the same results. However, the reciprocity-based approach provides a more general decomposition, applicable not only to the cases where the freedom and constraint subspaces do not span a six-dimensional space or have nonzero intersections, but also to the case where a manipulator has less than six DOFs [18].

If the specified twist and wrench are consistent with the environment (i.e. the freedom and constraint equations are satisfied), the specified task is feasible for hybrid control. In the opposite case, it is necessary to filter the specified twist and wrench to obtain a kinestatically realizable control action (so-called kinestatic filtering).

A procedure for applying the reciprocity-based task decomposition to manipulator dynamics in order to obtain motion equations relevant to hybrid control was presented in [18]. Several model-based tools for task specification using this approach are presented in [19]. The reciprocity concept not only has been shown to be well-suited to the nominal specification of arbitrary motion constraints, but also serves the definition of possible uncertainties and on-line identification and observation of real motion constraints [20]. This strategy makes task execution generally very robust against uncertainties. In particular, it is essential for contour following tasks. An overall hybrid position/force control scheme based on the general decomposition formalism, including identification of geometrical uncertainties, was proposed in [19].

The reason why almost all the above force control methods that are based on explicit force control cannot be applied with commercial robotic systems lies in the fact that commercial robots are designed as "positioning devices". In the above methods, the feedback term, i.e. the signal proportional to force errors, is multiplied by the transpose of the Jacobian matrix in order to calculate driving torques that have to be realized around the joints in order to achieve the desired force action. These signals are directly fed to the inputs of the local servo parts. However, the realization of computed torques by commercial robotic systems is not accurate, because the local servos are position-controlled and there is no force feedback with respect to the torques around the robot's joints. Consequently, the realization of desired torques is poor, since high friction and other non-linearities in the transmission mechanisms contribute significantly to the inaccuracy of current/torque causality. Besides, as there is no position feedback loop in the force-controlled direction, the robots will move due to various disturbances acting upon it, such as controller- and sensor-drifts, etc. [21]. The implementation of explicit force control in the presence of the above difficulties can be successful only with the new generation of direct-drive robots [22] with accurate joint torque control. Experimental evaluation and comparison of explicit force control strategies have been recently presented in [23].

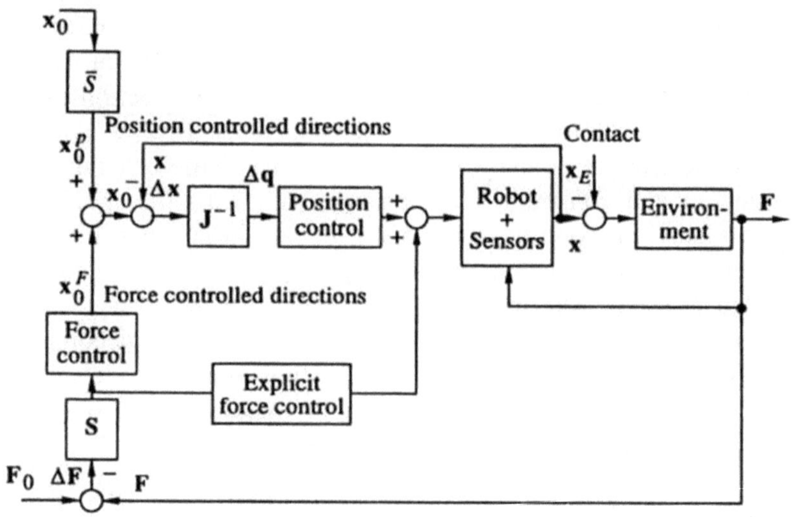

Figure A.1.3. Implicit hybrid position/force control.

With commercial applied robotic systems, it is most promising to implement *implicit* or *position-based* force control by closing a force-sensing loop around the position controller (Fig. A.1.3). The input to the force controller is the difference between the desired and actual contact force in the task frame. The output is an equivalent position in force-controlled directions which is used as a reference input to the positional controller. According to the hybrid force/position control concept, the equivalent position in the force direction x_0^F is superimposed to the orthogonal vector x_0^p in the task frame, which defines the nominal position in orthogonal position-controlled directions. The robot's behavior in the force direction is practically only affected by the acting force. The positional controller remains unchanged, except for the additional transformations between Cartesian and task frame, which have to be introduced since in a general case these two frames are not coincident. Since a position controller provides a basis for the realization of force control this concept is referred to as *implicit-* or *position-based* force control [24], or *external force* control [25].

A practical method for integrating implicit force control and implementing orthogonal decomposition in a position-controlled system for robot manufacturing applications (e.g., deburring) has been proposed in [26]. In the proposed scheme, the hybrid control integration is realized at the motion interpolation level (Fig. A.1.4). In order to decompose position- and force-controlled DOFs, the involvement of two virtual kinematic structures in Cartesian space is assumed: the first one is driven by the position interpolator, and the second structure receives the position commands (i.e. relative modifications) from the force control in the compliance (task) frame TF. The interpolator (IPO) generates the nominal position of the tool-center-point (TCP) in the position control DOFs in the task frame. The integral implicit force control modifies position of TF in force-controlled directions relative to its initial location with regard to the robot reference frame (e.g., robot base frame, object frame) according to the force error. By superposing the two positions (Fig. A.1.3), the command for internal position controller x_0 is obtained. In this way, independently of modifications of the tool position and/or orientation in the force direction (e.g., tool feed direction) according to interaction forces, it is possible to realize the desired motion in the position-controlled directions, so as to change tool clearance in the tool directions, etc.

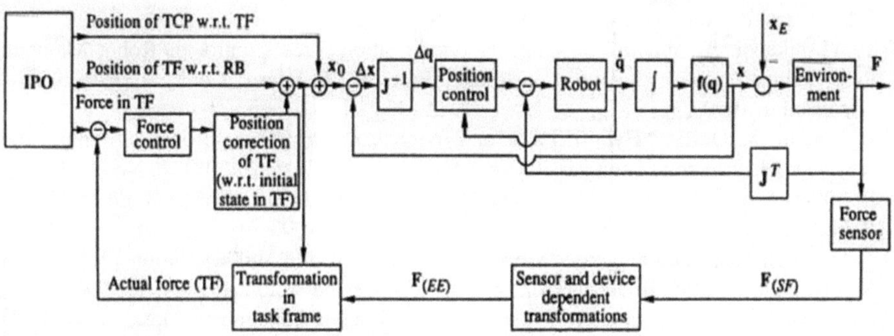

Figure A.1.4. Integration of implicit hybrid control.

REFERENCES

1. M.T. Mason, "Compliance and Force Control for Computer-Controlled Manipulators", IEEE Transactions on Systems, Man, and Cybernetics, Vol. SMC-11, No. 4, 1981.
2. R.P. Paul, B. Shimano, "Compliance and Control", Proc. Joint Automatic Control Conf., San Francisco, 1976.
3. M.H. Raibert, J.J. Craig, "Hybrid Position/Force Control of Manipulators", ASME J. Dynamic Systems, Measurement, and Control, Vol. 102, No. 2, 1981.
4. M. Vukobratovic, D. Surdilovic, "Control of Robotic Systems in Contact Tasks – An Overview", Izvestya of Russian Academy of Sciences: Theory and Control Systems, No. 5, Moscow, 1997.
5. D. Surdilovic, M. Vukobratovic, Control of Robotic Systems in Contact Tasks, chapter No. 23 in the Mechanical System Design Handbook – Modeling, Measurement and Control, Eds. Osita D. I. Nwokah, Yildirim Hurmuzlu, CRC Press, 2002.

6. C.H. An, J. Hollerbach, "Kinematic Stability Issues in Force Control of Manipulators", Proc. IEEE Intern. Conf. on Robotics and Automation, Raleigh, 1987.
7. H. Zhang, "A Kinematically Stable Hybrid Control Formulation", Proc. 3rd Intern. Symp. ISRAM, British Columbia, 1990.
8. W.D. Fisher, M.S. Mujtaba, "Hybrid Position/Force Control: A Correct Formulation", The Intern. J. Robotics Research, Vol. 11, No. 4, 1992.
9. C.H. An, J. Hollerbach, "Dynamic Stability Issues in Force Control of Manipulators", Proc. IEEE Intern. Conf. on Robotics and Automation, Raleigh, 1987.
10. J.Y.S. Luh, W.D. Fisher, R.P.C. Paul, "Joint Torque Control by a Direct Feedback for Industrial Robots", IEEE Transactions on Automatic Control, Vol. 28, No. 2, 1983.
11. K.G. Shin, C.P. Lee, "Compliant Control of Robotic Manipulators with Resolved Acceleration", Proc. 24th IEEE Conf. Decision and Control, 1985.
12. O. Khatib, "A Unified Approach for Motion and Force Control of Robot Manipulators: The Operational Space Formulation", IEEE J. Robotics and Automation, Vol. 3, No. 1, 1987.
13. B. Bona, M. Idri, "Exact Decoupling of the Force/Position Control Using the Operational Space Formulation", Proc. IEEE Intern. Conf. on Robotics and Automation, Nice, 1992.
14. T. Yoshikawa, M. Sugie, M. Tanaka, "Dynamic Hybrid Position/Force Control of Robot Manipulators Controller Design and Experiments", IEEE J. Robotics and Automation, Vol. 4, No. 6, 1988.
15. H. McClamroch, D. Wang, "Feedback Stabilization and Tracking of Constrained Robots", IEEE Trans. on Automatic Control, Vol. 33, No. 75, 1988.
16. T. Yoshikawa, A. Sudou, "Dynamic Hybrid Position/Force Control of Robot Manipulators: Estimation of Unknown Constraint", Proc. IEEE Intern. Conf. on Robotics and Automation, 1990.
17. H. Lipkin, J. Duffy, "Hybrid Twist and Wrench Control for a Robotic Manipulator", ASME J. Mechanisms, Transmission, and Automation in Design, Vol. 110, No. 2, 1988.
18. P. Sinha, A.A. Goldenberg, "A Unified Theory for Hybrid Control of Manipulators", Proc. IEEE Intern. Conf. on Robotics and Automation, Atlanta, 1993.
19. J. De Shutter, H. Bruyninckx, "Model-Based Specification and Execution of Compliant Motion", Tutorial M6, IEEE Intern. Conf. on Robotics and Automation, Nice, 1992.
20. H. Bruyninckx, J. De Schutter, B. Allota, "Model-Based Constrained Motion: Introducing On-Line Identification and Observations of Motion Constraints", Proc. 5th Intern. Conf. on Advanced Robotics, Pisa, 1991.
21. D.M. Stokic, D.T. Surdilovic, Position/Force Control of Industrial Robots: Implementation Problems, IPK Report, Berlin, IPK, 1990.
22. D. Visher, O. Khatib, "Design and Development of Torque-Controlled Joints", Proc. 1st Intern. Symp. on Experimental Robotics, 1989.
23. R. Volpe, P. Khosla, "An Experimental Evaluation and Comparison of Explicit Control Strategies for Robotic Manipulators", Proc. IEEE Intern. Conf. on Robotics and Automation, Nice, 1992.
24. J.A. Maples, J.J. Becker, "Experiments in Force Control of Robotics Manipulators", Proc. IEEE Intern. Conf. on Robotics and Automation, 1986.
25. J. De Schutter, H. Van Brussel, "Compliant Robot Motion II: A Control Based on External Control Loops", The Intern. J. Robotics Research, Vol. 7, No. 4, 1988.
26. G. Duelen, H. Munch, D. Surdilovic, et al. "Automated Force Control Schemes for Robotic Deburring Development and Experimental Evaluation", Proc. IECON, Special Session on Robotic Deburring, San Diego, 1992.

APPENDIX 2

SYNTHESIS OF CONTROL LAWS FOR THE ROBOT INTERACTING WITH DYNAMIC ENVIRONMENT

The model of dynamics of a robot interacting with the environment is described by vector differential nonlinear equations of the form:

$$H(q)\ddot{q}+h(q,\dot{q})=\tau+J^T(q)F \tag{A.2.1}$$

where, $q=q(t)$ is the n-dimensional vector of the robot generalized joint coordinates; $H(q)$ is the $n\times n$ positive definite matrix of inertia moments of the manipulation mechanisms; $h(q,\dot{q})$ is the n-dimensional nonlinear function of centrifugal, Coriolis' and gravitational moments; $\tau=\tau(t)$ is the n-dimensional vector of the input control; $J^T(q)$ is the $n\times m$ Jacobian matrix connecting velocities of the robot end-effector and velocities of the robot generalized coordinates; $F=F(t)$ is the m-dimensional vector of the generalized forces or generalized forces and moments acting on the end-effector from the environment. Dimension of the vector F can be adopted in Cartesian coordinates to be 3 or 6. Presently, it will be assumed that $n=m$ (generally $n > m$).

Because of the existence of the inverse of matrix $H(q)$, let us write the equation of robot dynamics model (A.2.1), solved with respect to the higher derivative, in the form:

$$\ddot{q}=-H^{-1}(q)h(q,\dot{q})+H^{-1}(q)\tau+H^{-1}(q)J^T(q)F \tag{A.2.2}$$

and, denote the right-hand side of equation (A.2.2) by $\Phi(q,\dot{q},\tau,F)$. Then, this equation can be written in the form:

$$\ddot{q}=\Phi(q,\dot{q},\tau,F),\quad q(t_o)=q_o,\quad \dot{q}(t_o)=\dot{q}_o \tag{A.2.3}$$

It is assumed that the function Φ satisfies all the conditions needed for the existence and uniqueness of the solution of the system of differential equations (A.2.3) on $[t_o, +\infty)$.

The robot dynamics model (A.2.1) can also be solved with respect to the control variable τ:

$$\tau=U(q,\dot{q},\ddot{q},F) \tag{A.2.4}$$

Let us now consider the mathematical model of the environment. It represents one of the most complex and delicate problems in the contact task control.

As we have seen, to describe environment we should generally use a different set of generalized coordinates, say q_e, when the environment has autonomous DOFs.

In the case when the environment does not possess displacements that are independent of the robot motion (the environment is firmly attached to the robot's end-effector, $s=\varphi(q)$), the environment dynamics can be described by the nonlinear differential equations*:

$$\begin{aligned} &M(s)\ddot{s}+L(s,\dot{s})=F \\ &s=\varphi(q) \end{aligned} \tag{A.2.5}$$

or in the frame of generalized coordinates of the robot (joint variables):

$$M(q)\ddot{q}+L(q,\dot{q})=S^T(q)F \tag{A.2.6}$$

where $M(q)$ is a positive definite $n\times n$ matrix; $L(q,\dot{q})$ is a nonlinear n-dimensional vector function; $S^T(q)$ is the $n\times n$ matrix with rank n, i.e. rank $(S)=n$. We also assume that all the mentioned matrices and vectors are continuous functions of their arguments.

By using the inverse matrix M, equation (A.2.6) can be written in the form:

* To make a difference, instead of the qecoordinates, when the environment possesses autonomous DOFs, we introduced S coordinates of the passive environment.

$$\ddot{q}=\Psi(q,\dot{q},F) \tag{A.2.7}$$

where,

$$\Psi(q,\dot{q},F)=-M^{-1}(q)L(q,\dot{q})+M^{-1}(q)S^{T}(q)F \ .$$

We assume that the vector function Ψ satisfies all the conditions for the existence and uniqueness of the Caushy problem solution for the system of differential equation (A.2.7) on $[t_o,+\infty)$.

Let us solve (A.2.6) with respect to the force F. Due to the fact that rank $(S)=n$, we have:

$$F=(S^{T}(q))^{-1}\left[M(q)\ddot{q}+L(q,\dot{q})\right]$$

By denoting the right-hand side of this equation by $f(q,\dot{q},\ddot{q})$, this can be written in the form:

$$F=f(q,\dot{q},\ddot{q}) \tag{A.2.8}$$

where f is a continuous function of its arguments.

The introduction of any force of action $F(t)$ on the environment will cause a unique motion of the robot $q(t)$ in that environment. The solvability property of the model equation (A.2.6) with respect to force action presumes that for the defined robot's motion $q(t)$ in the environment, the force:

$$F(t)\equiv f(q(t),\dot{q}(t),\ddot{q}(t))$$

is the force causing that motion.

In the case of contact with environment the robot control task can be described as the robot's motion along a programmed (desired) force of interaction, $F_p(t)$ acts between the robot and the environment. The desired programmed motion $(PM)q_p(t)$ and desired programmed interaction force $(PFI)F_p(t)$ cannot be arbitrary. The two functions must satisfy the following relation:

$$F_p(t)\equiv f(q_p(t),\ \dot{q}_p(t),\ \ddot{q}_p(t)) \tag{A.2.9}$$

The goal of the control of a robot interacting with dynamic environment can be formulated in the following way. Let us define the control $\tau(t)$ for $t\geq t_o$, to satisfy the target conditions:

$$q(t) \to q_p(t) \text{ as } t \to \infty, \quad F(t) \to F_p(t) \text{ as } t \to \infty \qquad \text{(A.2.10)}$$

A.2.1 Synthesis of Control Laws Stabilizing the Motion with the Preset Quality of Transient Processes

Let us describe the deviation of the real motion $q(t)$ from the programmed one by the transient response:

$$\eta(t) = q(t) - q_p(t)$$

It is natural to determine in advance the requirements to be satisfied by the transient response. These requirements enable certain quality of transient responses to be set in advance.

As some initial perturbations are always present, i.e. $\eta_o = \eta(t_o) = q(t_o) - q_p(t_o) \neq 0$, these requirements should not refer to a particular function $\eta(t)$, but to a whole family of functions $\{\eta(t)\}$, each of which acquires its concrete value in the initial moment. Therefore, the family of desired transient responses can be given by the vector differential equations: [1].

$$\ddot{\eta} = P(\eta, \dot{\eta}) \qquad \text{(A.2.11)}$$

where P is an n-dimensional vector function, continuous over the whole set of arguments, so that equation (A.2.11) has a unique trivial solution $\eta(t) \equiv 0$. Then, the fulfilment of the requirements that have been imposed on the transient processes is ensured by the choice of function P, and we shall design the robot control in such a way that in the absence of any perturbations apart from the initial ones, the robotic dynamics equation coincides with equation (A.2.11). We shall *a priori* adopt that the choice of the function P in the system (A.2.11) ensures asymptotic stability in the whole of trivial solution of the system (A.2.11).

As the function P can be adopted, for example, in the form: $P(\eta, \dot{\eta}) = \Gamma_1 \dot{\eta} + \Gamma_2 \eta$, then the system (A.2.11) obtains the form:

$$\ddot{\eta} = \Gamma_1 \dot{\eta} + \Gamma_2 \eta \qquad \text{(A.2.12)}$$

where Γ_1 and Γ_2 are the constant $n \times n$ matrices.

Let us turn from the equation (A.2.12) to a system of $2n$-order differential equations of the form [2]:

$$\dot{x}=\Gamma x \tag{A.2.13}$$

where :

$$x=(\eta,\dot{\eta})^T, \quad \Gamma=\begin{bmatrix} O_n & I_n \\ \Gamma_2 & \Gamma_1 \end{bmatrix}$$

O_n and I_n are the respective zero and unit matrices of dimension $n\times n$. Then, the matrices Γ_1 and Γ_2 can be chosen in such a way that the eigenvalues of the matrix Γ possess negative real parts. In this way, asymptotic stability as a whole of the solution of equation (A.2.13) and thus of equation (A.2.12), is achieved.

The task of stabilizing $(PFI)\, F_p(t)$ can be posed in an analogous way, by considering the family of transient responses with respect to force in the form [1, 2].

$$\dot{\mu}(t)=Q(\mu), \quad \mu(t)=F(t)-F_p(t) \tag{A.2.14}$$

and by choosing a continuous vector function $Q(Q(0)=0)$ of dimension n such that the asymptotic stability of the trivial solution of $\mu(t)\equiv 0$ as a whole is ensured.

At this point, a crucial question arises as to whether one can generally achieve that both the perturbed robot's motion and perturbed robot's interaction with the environment satisfy simultaneously equations (A.2.11) and (A.2.14). The answer is negative.

Let us explain this statement. Let $\tau=\tau(t),\, t\leq t_o$ be a control law, such that the robot's dynamics equation (A.2.1) with the control law is equivalent to the reference equation (A.2.11). Then, it follows from (A.2.11):

$$\ddot{q}=\ddot{q}_p+P(\eta,\dot{\eta}) \tag{A.2.15}$$

On the other hand, by comparing (A.2.8) and (A.2.9), we obtain the following relation between the real and programmed force of interaction, irrespective of the control law:

$$F-F_p=f(q,\dot{q},\ddot{q})-f(q_p,\dot{q}_p,\ddot{q}_p) \tag{A.2.16}$$

or, by taking into account (A.2.15):

$$\mu(t) = f(\eta + q_p,\ \dot{\eta} + \dot{q}_p, \ddot{q}_p + P(\eta, \dot{\eta})) - f(q_p, \dot{q}_p, \ddot{q}_p) \tag{A.2.17}$$

This shows that the character of change of $\mu(t)$ is fully defined by the right-hand side of (A.2.17) and depends on the function f determining the model of the external environment dynamics, the function P determining the transient response character of the robot's motion and the control τ via q and $\dot{q}$. In this way, $\mu(t)$ is actually a function which cannot be arbitrary and which, being the solution of equation (A.2.14), is dependent on the function Q that we have chosen in advance, and which is relatively arbitrary. This means that the quality of transient responses is controlled only in an implicit way via the function P, and is strongly dependent on the environment model.

Let us formulate two alternative questions:

- Can we choose such a control law that would enable the attainment of the control goals satisfying relation (A.2.10) by satisfying the preset robot's motion quality defined by (A.2.11)?
- Is it possible to choose the control law in such a way which will ensure the preset quality of the interaction force of the robot interacting with environment, defined by (A.2.14), and also the attainment of the control goals?

If the answer to the first question is positive (which will be discussed in the following text), the answer to the second question depends on the environment dynamics model, and this will be considered in the next section.

Let us synthesize the control law $\tau(t)$ in such a way as to ensure the desired quality of robot's motion (A.2.11).

Then the relation (A.2.15) has to be satisfied in the closed loop. This can be achieved by adopting the control law that has feedback loops with respect to $q, \dot{q}$ and F [3,4]:

$$\tau = H(q)\left(\ddot{q}_p + P(\eta, \dot{\eta})\right) + h(q, \dot{q}) - J^T(q)F \tag{A.2.18}$$

By substituting (A.2.18) into the robot's dynamics model (A.2.1) we obtain equation of the closed-form system:

$$H(q)\left(\ddot{q} - \ddot{q}_p - P(\eta, \dot{\eta})\right) = 0 \tag{A.2.19}$$

Because $H(q)$ is a positive definite matrix, $\det H \neq 0$, and consequently, (A.2.19) is equivalent to (A.2.11). Due to the property of the function P, which ensures asymptotic stability of the system (A.2.11) as a whole with respect to $(\eta, \dot{\eta})$, we will have:

$$\eta(t)_{t\to\infty} \to 0 \quad \text{and} \quad \dot{\eta}(t)_{t\to\infty} \to 0 \tag{A.2.20}$$

and, since P is a continuous function and $P(0,0)=0$, it follows:

$$\ddot{\eta}(t)_{t\to\infty} \to 0 \tag{A.2.21}$$

By taking into account the previously obtained relations (A.2.16) or (A.2.17), and also the continuity of the function f, we obtain:

$$\mu(t)_{t\to\infty} \to 0 \tag{A.2.22}$$

Thus, the control goal is achieved, i.e. along with the closeness of the real and programmed robot's motion, the closeness of the real $F(t)$ and programmed $F_p(t)$ force of interaction between the robot and environment is attained.

Let us consider one more control law of the robot motion stabilization without utilizing force feedback, when there exists an interaction force between the robot and environment. For this purpose, let us modify the control law (A.2.18) using relations (A.2.8) and (A.2.15). We obtain:

$$\tau = H(q)\left[\ddot{q}_p + P(\eta,\dot{\eta})\right] + h(q,\dot{q}) - J^T(q) f(q,\dot{q},\ddot{q}_p + P(\eta,\dot{\eta})) \tag{A.2.23}$$

Let us show that the control law (A.2.23) with feedback loops only with respect to q and $\dot{q}$ solves the task of stabilizing *PM* (A.2.20), (A.2.21), and consequently, in accordance with (A.2.22), the control task itself under conditions of the robot's contact with environment, together with the supplementary assumption.

By introducing the control (A.2.23) into the robot dynamics model (A.2.1) we obtain:

$$H(q)\left[\ddot{\eta} - P(\eta,\dot{\eta})\right] = J^T(q)(F - f(q,\dot{q},\ddot{q}_p + P(\eta,\dot{\eta})) \tag{A.2.24}$$

Taking into account the equality (A.2.8), we have:

$$H(q)\left[\ddot{\eta} - P(\eta,\dot{\eta})\right] = J^T(q)(f(q,\dot{q},\ddot{q}) - f(q,\dot{q},\ddot{q}_p + P(\eta,\dot{\eta})) \tag{A.2.25}$$

or, by taking into account the form of the function f :

$$H(q)\left[\ddot{\eta} - P(\eta,\dot{\eta})\right] = J^T(q)(S^T(q))^{-1} M(q)\left[\ddot{\eta} - P(\eta,\dot{\eta})\right] \tag{A.2.26}$$

Thereby

$$(H(q)-J^T(q)(S^T(q))^{-1}M(q))[\ddot{\eta}-P(\eta,\dot{\eta})]=0 \tag{A.2.27}$$

Let us suppose that for all q, i.e. for an arbitrary robot configuration, the supplementary condition

$$\det\left(H(q)-J^T(q)(S^T(q))^{-1}M(q)\right)\neq 0 \tag{A.2.28}$$

is fulfilled. Then, the equation of closed-loop system (A.2.27) is equivalent to the reference equation (A.2.11).

Although the control law (A.2.23) does not use force feedback, it is explicitly dependent on the environment dynamics:

$$\tau=U(q,\dot{q},\ddot{q}_p+P(\eta,\dot{\eta}),\ f(q,\dot{q},\ddot{q}_p+P(\eta,\dot{\eta}))) \tag{A.2.29}$$

through the function f, and cannot be used to control the manipulator in free space.

A2.2 Synthesis of Control Laws Stabilizing the Interaction Force of the Robot and the Environment with the Preset Quality of Transient Processes

As shown in the preceding section, the answer to the first question was quite simple: the inverse dynamics method ensures achievement of a desired motion quality and at the same time guarantees stability of the interaction force.

Let us now consider the issue of stabilizing the force of interaction of the robot with environment. Let us first discuss a few control laws ensuring solution of this task. For convenience, when describing the quality of transient responses with respect to force (A.2.14), we will use an equivalent relation of the form [1,2]:

$$\mu(t)=\mu_o+\int_{t_o}^{t}Q(\mu(\omega))d\omega \tag{A.2.30}$$

With no loss in generality we can adopt $\mu\equiv 0$, because the stabilization of μ in the sense of preset quality (A.2.30) directs stabilization according to the preset quality (A.2.14).

Let us consider the control law with feedback with respect to $q, \dot{q}$ and F, of the form:

$$\tau = H(q)M^{-1}(q)\left(-L(q,\dot{q}) + S^T(q)(F_p + \int_{t_0}^{t} Q(\mu(\omega))d\omega) \right) + \qquad \text{(A.2.31)}$$
$$+ h(q,\dot{q}) - J^T(q)F$$

By introducing this control law into the robot's dynamics model (A.2.1) we obtain the following law of robot's functioning in contact with the environment:

$$M(q)\ddot{q} + L(q,\dot{q}) = S^T(q)(F_p + \int_{t_o}^{t} Q(\mu(\omega))d\omega) \qquad \text{(A.2.32)}$$

Taking into account the environment dynamics model (A.2.1) we obtain the following closed-loop control system:

$$S^T(q)(\mu(t) - \int_{t_o}^{t} Q(\mu(\omega))\, d\omega) = 0$$

and, because rank $(S) = n$, this equation is equivalent to the equation:

$$\mu(t) = \int_{t_o}^{t} Q(\mu(\omega))\, d\omega \qquad \text{(A.2.33)}$$

whereby (A.2.14) follows straight away.

In this way the control law (A.2.31) ensures a desired quality of stabilization of $(PFI)\, F_p(t)$. Let us note that the control law (A.2.31) can be rewritten in a more compact form as:

$$\tau = U(q,\dot{q},\psi(q,\dot{q},F_p + \int_{t_o}^{t} Q(\mu(\omega))\, d\omega), F) \qquad \text{(A.2.34)}$$

Let us consider another control law of the form

$$\tau = U(q,\dot{q},\psi(q,\dot{q},F),\ F_p + \int_{t_o}^{t} Q(\mu(\omega))\, d\omega) \qquad \text{(A.2.35)}$$

involving feedback loops with respect to $q, \dot{q}$ and F, and let us demonstrate that it also ensures a desired quality of stabilizing $(PFI)\,F_p(t)$. In a developed form, this control law can be written as:

$$\tau = H(q)M^{-1}(q)\left(-L(q,\dot{q}) + S^T(q)F\right) + h(q,\dot{q}) - \\ - J^T(q)(F_p + \int_{t_o}^{t} Q(\mu(\omega))d\omega) \tag{A.2.36}$$

By substituting τ from (A.2.36) into (A.2.1) we obtain the closed-form equation:

$$H(q)\ddot{q} = H(q)M^{-1}(q)\left(-L(q,\dot{q}) + S^T(q)F\right) + \\ + J^T(q)(\mu - \int_{t_o}^{t} Q(\mu(\omega))d\omega) \tag{A.2.37}$$

Taking into account that for the contact of the robot with environment holds equation (A.2.6), i.e.

$$-L(q,\dot{q}) + S^T(q)F \equiv M(q)\ddot{q}$$

relation (A.2.37) is equivalent to the relation:

$$J^T(q)(\mu - \int_{t_o}^{t} Q(\mu(\omega))\,d\omega) = 0$$

Assuming that rank $J(q) = n$ for all robot configurations, we obtain a closed-form equation of the system in the form (A.2.30).

The requirement rank $J(q) = n$ is quite natural if it is borne in mind that from the point of view of physics this means that the magnitude of the force of interaction of robot (A.2.1) with its environment (A.2.6) is uniquely determined by the control and motion of the robot. Moreover, it can be stated that in the case of robot's contact with environment no configurations for which rank $J(q) < n$ could arise, as this would mean that the force interaction satisfying the equation of environment's dynamics (A.2.8) would not be uniquely determined by the motion law $q(t)$.

Finally, let us consider one more control law stabilizing $(PFI)\,F_p(t)$ with a preset quality of transient process, using no explicit force feedback but the implicit one, via the integral of the difference $F(t) - F_p(t)$.

Let the control law be given in the form:

$$\tau=U(q,\dot{q},\psi(q,\dot{q},F_p+\int_{t_o}^{t}Q(\mu(\omega))d\omega),\ F_p+\int_{t_o}^{t}Q(\mu(\omega))\,d\omega)\quad \text{(A.2.38)}$$

or, in a more developed form, as

$$\tau=HM^{-1}\left(-L+S^T(F_p+\int_{t_o}^{t}Q(\mu)d\omega)\right)+h-J^T(F_p+\int_{t_o}^{t}Q(\mu)d\omega).$$

The closed form of the control system is then

$$H\ddot{q}=HM^{-1}\left(-L+S^T(F_p+\int_{t_o}^{t}Q(\mu)d\omega)\right)+ +J^T(\mu-\int_{t_o}^{t}Q(\mu)d\omega)\quad \text{(A.2.39)}$$

By eliminating $\ddot{q}$ from equations (A.2.39) and (A.2.7) we obtain

$$-L+S^TF=-L+S^T(F_p+\int_{t_o}^{t}Q(\mu)d\omega)+MH^{-1}J^T(\mu-\int_{t_o}^{t}Q(\mu)d\omega)$$

or $S^T(\mu-\int_{t_o}^{t}Q(\mu)d\omega)=MH^{-1}J^T(\mu-\int_{t_o}^{t}Q(\mu)d\omega)$

Thereby,

$$(S^T(q)-MH^{-1}J^T(q))(\mu-\int_{t_o}^{t}Q(\mu)d\omega)=0\quad \text{(A.2.40)}$$

Let us set the supplementary requirement as

$$rank\left[S^T(q)-M(q)H^{-1}(q)J^T(q)\right]=n\quad \text{(A.2.41)}$$

for all possible robot configurations. Then, relation (A.2.40) will be identical to (A.2.14) and, consequently, under the supplementary assumption (A.2.41), the control law (A.2.34) also ensures stabilization of $(PFI)\,F_p(t)$ with a preset quality of transient regimes.

It is easy to notice that because of the continuity of robot's motion, the violation of the conditions (A.2.41) over a finite or countable set of points does not entail violation of the control system functioning in a feedback in the form (A.2.33). The violation of this condition may generally occur only if the inequality $rank\left[S^T(q) - M(q)H^{-1}(q)J^T(q)\right] < n$ has over any time interval twice continuously differentiable solution $q(t),\, t \in (t_a, t_b),\, a < b$. This question requires additional investigations of dynamics models of the robot and environment. However, attention has to be paid to the possibility of regimes appearing analogous to sliding regimes in the tasks of dynamic objects control $[4,6]$.

So, we have synthesized three control laws (A.2.34), (A.2.35), and (A.2.38) each of which ensures a desired quality of stabilization of the $(PFI)\, F_p(t)$.

REFERENCES

1. M. Vukobratovic, and Yu. Ekalo, "New Approach to Control Laws Synthesis for Robotic Manipulators in Contact with Dynamic Environment", Tutorial S5: Force and Contact Control in Robotic Systems: A Historical Perspective and Current Technologies, in: IEEE Int. Conf. on Robotics and Automation, Atlanta, pp. 213-229, 1993.
2. M. Vukobratovic, and Yu. Ekalo, "New Approach to Control of Robotic Manipulators Interacting with Dynamic Environment", Robotica 14, pp. 31-39, 1996.
3. Yu. Ekalo and M. Vukobratovic, "Quality of Stabilization of Robot Interacting with Dynamic Environment", Journal of Intelligent and Robotic Systems 14(2), pp. 155 -179, 1995.
4. M. Vukobratovic, "The Role of Environment Dynamics in Contact Force Control of Manipulation Robots", ASME Journal of Dyn. Systems, Measurements and Control, March issue, 1997.
5. M. Vukobratovic, A. Rodic, and Yu. Ekalo, "Impedance Control as a Particular Case of the Unified Approach to the Control of Robots Interacting with a Dynamic Environment", Journal of Intelligent and Robotic Systems, Vol 18, pp. 191-204, 1997.
6. M. Vukobratovic, R. Stojic, Yu. Ekalo, "Contribution to the Problem Solution of Position/Force Control of Manipulation Robots in Contact with Dynamic Environments – A Generalization", IFAC Automatica, Vol. 34, No 10, 1998.

Index

MIX
Papier aus verantwortungsvollen Quellen
Paper from responsible sources
FSC® C105338

If you have any concerns about our products,
you can contact us on
ProductSafety@springernature.com

In case Publisher is established outside the EU,
the EU authorized representative is:
Springer Nature Customer Service Center GmbH
Europaplatz 3, 69115 Heidelberg, Germany

Printed by Libri Plureos GmbH
in Hamburg, Germany